Managing Postharvest Quality and Losses in Horticultural Crops

The Editors

Dr. K.L. Chadha is currently the President of The Horticultural Society of India as also Adjunct Professor (Hort.) at IARI. He received his B.Sc. (Agri) and M.Sc. (Hort.) degrees from Punjab University, and Ph.D. from Indian Agricultural Research Institute (IARI), New Delhi.

He has served in various capacities at IARI, Punjab Agricultural University, Ludhiana Central Mango Research Station, Lucknow. He was the Director, IIHR, Bangalore, Horticulture Commissioner and Executive Director, National Horticulture Board, Govt. of India; Deputy Director General (Hort.), ICAR and National Professor (Hort.) at IARI.

He has authored/edited nearly 4 dozen books and bulletins including the series "Advances in Horticulture" published in 13 volumes covering 9,410 pages. He has been associated with several scientific societies as their Chief Patron, President or Vice President. He has served as consultant to several national and international organisations.

Dr. Chadha is the recipient of several awards for his outstanding contributions to R & D in horticulture. He is a Fellow of NAAS, New Delhi and ISHS, Belgium and D.Sc. (h.c.) from three State Agricultural Universities. Some important awards won by him include Borlaug Award, Om Prakash Bhasin Award, Dr. H.M. Marigowda Award, S.K. Mitra Memorial Award, S.S. Ranade Memorial Life Time Achievement Award, HSI-Shivashakthi Life-Time Achievement Award, B.P. Pal Memorial Award, National Agriculture Leadership Award and Krishi Shiromanni Samman.

Dr K.L. Chadha has steered policy planning, research and development of horticulture in India for almost five decades. For his yeomen's service Dr Chadha was conferred the prestigious *Padma Shri* by Her Excellency, the President of India.

Prof. R. K. Pal is currently working as the Director of the ICAR-National Research Center on Pomegranate at Solapur, Maharashtra. After his formal education from West Bengal, he received M.Sc. and Ph.D. degrees in Horticulture from IARI and joined Agricultural Research Service of ICAR. He served IARI in different capacities for over 25 years as Scientist, Professor and Head of the Division of PHT. He underwent advanced training in USA and attended the International Course on Food Processing as FAO Fellow in the Netherlands. He has published more than 100 research/technical papers in national and international journals and guided several research scholars at IARI, New Delhi currently occupying prestigious positions in India, USA, Iran, Vietnam and in many other countries. He is a Fellow of Horticultural Society of India, and has won Dr. J.C. Anand Gold Medal for his contribution in PHT of horticultural crops. He was a national consultant of FAO, member of the Technical Committee of SAARC and Technical Consultant for FHEL-operated CA store. He represented IARI as a core committee member for establishment of Advance Centre for Agriculture Research and Education in Myanmar.

Managing Postharvest Quality and Losses in Horticultural Crops

Volume 3
Vegetables, Flowers & Plantation Crops

Editors

K.L. Chadha

President, The Horticulture Society of India,
F-1 Block, NASC Complex, DPS Marg, Todapur,
New Delhi – 110 012, India

and

R.K. Pal

Director, ICAR-National Research Centre on Pomegranate
NH-9, Solapur-Pune Highway, Kegaon,
Solapur – 413 255, Maharashtra, India

2019

Daya Publishing House®

A Division of

Astral International Pvt. Ltd.

New Delhi – 110 002

Managing Postharvest Quality and Losses in Horticultural Crops (3 Volume Set)

Volume 1: General Issues	Page 001-251
Volume 2: Fruit Crops	Page 253-579
Volume 3: Vegetables, Flowers & Plantation Crops	Page 581-755

First Published, 2015

ISBN 978-93-5124-357-1 (Vol.3)
ISBN 978-93-5124-358-8 (Set)

Published by : **Daya Publishing House®**
A Division of
Astral International Pvt. Ltd.
– ISO 9001:2015 Certified Company –
4736/23, Ansari Road, Darya Ganj
New Delhi-110 002
Ph. 011-43549197, 23278134
E-mail: info@astralint.com
Website: www.astralint.com

Laser Typesetting : **Classic Computer Services**, Delhi - 110 035

Printed at : **Replika Press Pvt. Ltd.**

PRINTED IN INDIA

Prof. M. S. Swaminathan
Founder Chairman, M S Swaminathan Research Foundation
Third Cross Street, Taramani Institutional Area
Chennai – 600 113 (India)
Email: swami@mssrf.res.in

MSSRF

Foreword

During 1981 the Government of India had constituted a committee under my chairmanship to suggest a scientific approach to postharvest management of perishables in the country. Among various recommendations made, the Committee suggested the establishment of a National Horticulture Board to lay emphasis on programmes related to Postharvest Management. The Committee also identified a group of 12 horticultural crops covering fruits, vegetables and spices for postharvest research apart from several other new crops that were gaining commercial importance. Among fruits apple, mango, banana, citrus, grape, guava, papaya, pineapple, pomegranate and sapota were identified. While among vegetables and spice crops commercial crops like onion, potato, pepper/chilli, turmeric and ginger were included. The National Horticulture Board of the Ministry of Agriculture with its headquarters at Gurgaon is an important outcome of the recommendations.

Even though research on Postharvest Management has been underway in India at the Central Food Technological Research Institute (CFTRI), Mysore, it was not adequate to cover all major crops and aspects. Thus ICAR as a nodal organization implemented a project on Postharvest Management under PL-480 scheme and subsequently set up the Central Institute for Post Harvest Engineering and Technology at Ludhiana. In addition, Defence Food Research Laboratory (DFRL), Indian Institute of Crop Processing Technologies (IICPT) Thanjavur, Indian Institute of Technology (IIT) Kharagpur and Mumbai, Jhadavpur University, Kolkata, Bhaba Atomic Research Centre (BARC), Mumbai and many other private funded Institutions have been giving emphasis on these aspects. Of late, new institutions *viz.* National Institute of Food Technology Entrepreneurship and Management (NIFTEM) at Kondili, Haryana and National Agri-Food Biotechnology Institute (NABI), Mohali, Punjab have also been established in view of the emergent need for combating the huge postharvest losses of perishable commodities taking place in India. As a result several technologies have been developed on postharvest management of perishable horticultural crops by these institutions. However, most of these technologies have not been transferred to the end users in an appropriate manner due to non-availability of recommendations in an implementable form. Further, all these recommendations are not available at one place and are mostly concealed in research papers or reports which are neither available nor comprehensive.

I am happy that Dr. K.L. Chadha, former Deputy Director General (Horticulture), ICAR and Dr. R.K. Pal, Director, ICAR-National Research Centre on Pomegranate, Solapur, Maharashtra have made an excellent effort in consolidating the scattered information available from various source, compiled it and edited it in a lucid and understandable format in this book titled "Managing Postharvest Quality and Losses in Horticultural Crops" with conclusive recommendations for the benefit of different stake holders. I congratulate the editors for this effort at a time when the horticulture production has overtaken the food crop production and its post-harvest management warrants the use of scientific technologies.

I hope this book will be of useful to all engaged in wholesale, retail and export trade of horticultural crops as well as researchers and students. It is also hoped that researchers will make further efforts to plug the gaps in technologies wherever these exist. At the same time, we should continue to make effort to bridge the gap between scientific know-how and field level do-how.

M.S. Swaminathan

Preface

India is the second largest producer of fruits and vegetables in the world but it encounters huge postharvest losses almost equivalent to the total production of horticultue crops of the United Kingdom. Reduction in postharvest losses of horticultural crops is less expensive than bringing equivalent increase in production. Postharvest research in India was started at CFTRI, Mysore. However, with the increase in production and productivity of horticultural crops in India, ICAR launched a PL-480 scheme with the assistance of the United States of America to undertake research on postharvest technology of horticultural crops at IARI, New Delhi as lead centre. Later on an All India Coordinated Research Project on Postharvest Technology of Horticultural Crops was launched with a few centres at ICAR institutes and state agricultural universities. During the last three decades or so, lot of information on several aspects of postharvest management of horticultural crops has been generated in India. The horticulture industry in India has grown leaps and bounds with the increase in area under cultivation, production and export. However, scienitific postharvest management and leading adoption of international standards for export quality of horticultural produce from India with respect to combating various tariff and non-tariff barriers is still in its infancy. This book has compiled information keeping in view the needs of horticultural entreprenerus who would like to take-up domestic and export trade of fruits and vegetables on scientific basis and do not have access to this vital information.

The book is divided into three volumes comprising of 1) General Issues, 2) Postharvest Management of Fruit Crops and 3) Postharvest Management of Vegetables, Flowers and Plantation Crops. Eminent research workers who have been engaged in the area of Postharvest Management in the NARS have contributed chapters in this publication. Efforts have been made to incorporate information on status of cultivation, cultivars, association between pre-harvest crop management practices and quality after harvest, pack-house operations, storage and marketing for domestic and export trade. The book is also suitably illustrated with colour photographs, wherever necessary. Important references to each subject have been cited for detailed reading. The authors acknowledge the compilation of chapters using scattered information that is available in public domain in the form of research papers and in relevent websites.

The editors are grateful to all the contributors for their cooperation, support and timely submission of their manuscripts for bringing out this publication. The editors also express thanks to Mrs. Mandira Pal for her active support in correcting proofs of the book. The cooperation and support of the staff of the Horticultural Society of India especially Mr. Santosh Kr. Atre is duly acknowledged. The editors are extremely grateful to Prof. M.S. Swaminathan for agreeing to write the Foreword of the book.

It is hoped that this publication will be of immense value to all the stakeholders *viz.* horticultural entrepreneurs, retailers, wholesalers and exporters engaged in the business of supply chain of perishable horticultural produce, research scholars engaged in development of postharvest management protocols, and others in view of the objectives for which it has been written, *i.e.,* reducing postharvest losses and delivering quality produce both in domestic and international market.

K.L. Chadha

R.K. Pal

Contents

Volume 2: Fruit Crops

Volume 3: Vegetables, Flowers and Plantation Crops

List of Contributors Vol. 3

Arulraj, S. (Dr.)
Director, ICAR-Directorate of Oil Palm Research, Pedavegi – 534 450, Andhra Pradesh, India
E-mail: dopr2009@gmail.com

Balasubramanian, D. (Dr.)
Principal Scientist (ASPE), ICAR-Directorate of Cashew Research, Puttur – 574 202, Dakshina Kannada, Karnataka, India
E-mail: bavika13@email.com

De, L.C. (Dr.)
Principal Scientist, ICAR-National Research Centre for Orchids, Pakyong – 737 106, Sikkim, India
E-mail: lcde_horti@rediffmail.com

Ezekiel, R. (Dr.)
Former Head, Division of Crop Physiology, Biochemistry and Postharvest Technology, ICAR-Central Potato Research Institute, Shimla – 171 001, H.P., India
E-mail: rajarathnamezekiel@yahoo.co.in

Gupta, R.P. (Dr.)
Director, National Horticultural Research and Development Foundation, Nashik – 422 003, Maharashtra, India, India
E-mail: nashik@nhrdf.com

Kalidasu, Giridhar (Dr.)
Scientist (Horticulture), AICRP on Spices, Dr. Y.S.R. Horticultural University, Horticultural Research Station, Lam, Guntur – 522 034, India
E-mail: gkalidasu@yahoo.com

Kumar, Ramesh (Dr.)
Director, Directorate of Floricultural Research, ICAR-Indian Agricultural Research Institute, New Delhi – 110 012, India
E-mail: directordfr@gmail.com

Lawande, K.E. (Dr.)
Vice Chancellor, Dr. Balasaheb Sawant Konkan Krishi Vidyapeeth, Tal. Dapoli, Dist. Ratnagiri – 415 712, Maharashtra, India
E-mail: kelawande2011@gmail.com

Pathak, N. (Dr.)
Sikkim Himalayan Orchids Ltd., Assam Lingzey – 737 135, East Sikkim, India
E-mail: narender.pathak@gmail.com

Saikia, Ananta (Dr.)
Professor, Department of Horticulture, Assam Agricultural University, Jorhat – 785 013, Assam, India
E-mail: ananta.saikia@gmail.com

Saroj, P.L. (Dr.)
Director, ICAR-Directorate of Cashew Research Puttur – 574 202, Dakshina Kannada, Karnataka, India
E-mail: plsaroj@yahoo.co.in

Sharma, H.P. (Dr.)
Scientist, National Horticultural Research and Development Foundation, Nashik – 422 003, Maharashtra, India
E-mail: nashik@nhrdf.com

Sharma, Sanjib (Dr.)
Assistant Professor, Department of Horticulture, Assam Agricultural University, Jorhat – 785 013, Assam, India
E-mail: sanjibsir@rediffmail.com

Singh, Kushal (Dr.)
Senior Plant Physiologist, Department of Floriculture and Landscaping, Punjab Agricultural University, Ludhiana – 141 004, Punjab, India
E-mail: kushal.flori@pau.edu

Surepeddi, Surya Kumari (Dr.)
Principal Scientist (Horticulture), AICRP on Spices, Dr. Y.S.R. Horticultural University, Horticultural Research Station, Lam, Guntur – 522 034, India
E-mail: surya.surepeddi@gmail.com

Tripathi, P.C. (Dr.)
Head, Central Horticultural Experiment Station (IIHR), Chettalli – 571 248, Karnataka, India
E-mail: prakaashtripathii2000@yahoo.co.in

Chapter 27
Cashew

D. Balasubramanian
Principal Scientist
P.L. Saroj
Director
ICAR-Directorate of Cashew Research,
Puttur – 574 202, Dakshina Kannada, Karnataka, India
E-mail: bavika13@email.com, E-mail: plsaroj@yahoo.co.in

1.0 Introduction

Cashew (*Anacardium occidentale*) is originally native to northeastern Brazil. It was introduced to the Malabar Coast of India in the sixteenth century by the Spaniards and probably served as a locus of dispersal to other centers in India and South East Asia. It is considered as an important plantation crop in India due to its utility in soil and water conservation and to build up of balanced ecosystem. Among horticultural crops, cashew has been known to provide very high economic returns. The cashew nut, popular dry fruit is sweet and delightful in taste. It is a valuable food for both physical and mental health. It is considered as complete nourishing diet and functional food for various ailments. The analysis of the nut shows that on an average it contains 21.2 per cent protein, 2.4 per cent mineral matter, 22.3 per cent carbohydrate, 46.9 per cent fat, 1.3 per cent fibre, 0.05 per cent, calcium, 0.45 phosphorous and 5.0 mg per 100 gm iron. It also contains carotene as vitamin A 100 IU, thiamine 630 mg, niacin 2.1 mg and riboflavin 190 mg per 100 gm of kernel. It is free from cholesterol and its calorific value is more than egg. The presence of vitamin E in cashew takes care of all reproductive problems and prevents the development of oxidative rancidity in fats. Recently discovered vitamin-PP also develops while roasting cashew and it can exert a sparing action on the B group vitamins and assist in metabolism of lactose and thiamine.

2.0 Status of Cultivation

In India, cashew is cultivated in a total area of 9.91 lakh ha with an annual production of 6.92 lakh MT during the year 2011-12. The average productivity in this country is around 749 kg ha^{-1} which is 30 per cent higher than productivity of cashew at global level. Significant increase in the production of cashew at global level registered over the past 50 years (Table 1). Africa was the largest producer of cashew till the 70s, accounting for about 68 per cent of world's cashew production. During this period, Mozambique and Tanzania were the dominant players in the market. In the last three decades, however, there has been a transformation in the global production scenario with Asia overtaking Africa. India till recently led the world production, but the production data available from Food and Agricultural Organization (FAO) indicated that Vietnam and Nigeria have outstripped India and Brazil in cashew production during the year 2011.

Table 1: Cashew Production in Major Cashew Growing Countries (1961-2011)

(Quantity in MT)

Country	*Year*					
	1961	*1971*	*1981*	*1991*	*2001*	*2011*
India	85000	127223	185250	294590	450000	674600
Brazil	9670	28602	74000	185965	124073	230785
Mozambique	107000	202000	71100	31134	58000	72263
Nigeria	7000	25000	25000	45000	485000	813023
Philippines	7500	5014	3786	98283	111153	133388
Vietnam	700	3500	6500	160000	292800	1272000
Tanzania	50000	126409	64000	29850	122000	75000
World	287535	558846	494157	1003094	2064375	4201010

Source: FAO, 2011.

In India, cashew is largely grown in different states of the coastal regions *viz.*, Kerala, Karnataka, Goa, Maharashtra, Tamil Nadu, Andhra Pradesh, Orissa, West Bengal and North East Hilly region (NEH) (Table 2). The Directorate of Cashew and Cocoa Development, the Directorate of Cashew Research, the Cashew Export Promotion Council and the Agricultural Universities across the states have been continuously intervening for the development of high yielding varieties, production of grafts and high density planting systems for augmenting the production of quality cashew nuts in this country. Current production data reveals that Maharashtra had emerged as the dominant player in the market contributing almost one third of the national production but accounting for only one fifth of the land area under cashew cultivation, showing a higher efficiency in production. Productivity is found to be nearly one and half times more than the national average.

Table 2: Area, Production and Productivity of Cashew in India (2011-12)

State	*Area ('000 ha)*	*Production ('000 MT)*	*Productivity (MT ha^{-1})*
Andhra Pradesh	184.00	110.00	0.60
Goa	58.00	25.00	0.46
Karnataka	121.00	60.00	0.52
Kerala	83.00	73.00	0.95
Maharashtra	183.00	223.00	1.18
Orissa	158.00	97.00	0.68
Tamil Nadu	136.00	68.00	0.52
West Bengal	11.00	10.00	0.91
Others*	48.00	31.00	0.86
Total	991.00	692.00	0.75

* Includes Jharkhand, Chhattisgarh, Meghalaya, Manipur, Assam, Tripura and Nagaland.

Source: Directorate of Cashew and Cocoa Development, Cochin, Kerala.

3.0 Commercial Varieties

a) Bhaskara

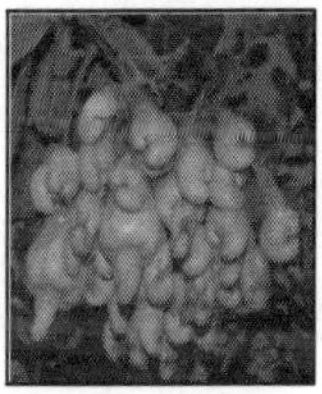

This variety was released during March 2006 for coastal region of Karnataka. This is having midseason flowering habit (Dec-Mar) with a flowering duration of 60 days and has potential to escape from the attack of the tea mosquito bug (TMB) under low to moderate outbreak situation. But the regular insecticidal spray against TMB is essential under severe out break situation. The number of fruits per panicle (bunch) ranged from 4 -13. The average yield on 13th year was 10.7 kg/tree with highest yield of 19 kg/tree. The nut and kernel weight are 7.4 g and 2.2 g respectively. The shelling percentage is 30.6 and kernel grade conforms to export grade W240. The apple colour is pinkish orange and juice content is 67.5 per cent.

b) BPP 8 (H 2/16)

It is a hybrid (H2/16) derived from the cross Tree No.1 x Tree No.39 and released in 1993 for general cultivation in Andhra Pradesh. It has been performing well in Orissa and West Bengal also. This variety is superior to all the other six varieties developed from Bapatla. The variety has mean yield of 14 kg/tree with better nut size (8.2g). Shelling percentage (29 per cent) of this variety is also better than the rest of the varieties released from Bapatla so far. Kernel grade is W 210 (export grade).

c) NDR 2-1 (Madakkathara-2)

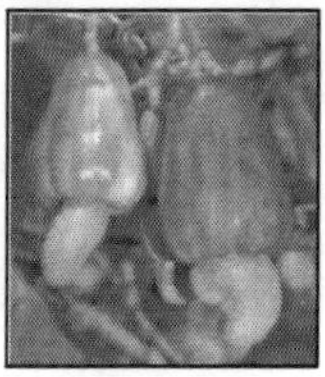

This is a selection from germplasm collection made from Neduvellur in Kerala maintained at CRS, Anakkayam. This variety was also released in 1987. The mean yield is 17 kg/tree. The nuts are bold (7.3 g nut weight) with shelling percentage of 26.2 per cent.

Kernel weight is 2g having a count of W 240 export grade. Apple colour is red and with weight of apple 63.3g. Reducing sugar 7.8 per cent.

d) Priyanka (H 1591)

This is a new hybrid with parentage of BLA 139-1 X K 30-1 with jumbo nut size developed and released from Cashew Research Station, Madakkathara in 1995. The yield of nuts is 16.9 kg/tree. The nut weight is 10.8g with kernel weight of 2.87g. Shelling percentage is 26.5. The export grade of kernels conforms to W 180. Colour of apple is yellowish red weighing 135g with juice content of 57.4 per cent.

e) Selection-2

This is a selection from the segregating seedling progeny of 2/9 Dicherla (VTH 40/1) originally a collection made from Andhra Pradesh. This variety was released in 1989. It has early season flowering habit (Nov - Jan) with a flowering duration of 74 days. The number of fruits per bunch is three with the yield of 9 kg/tree. The nut weight is 9.2g with shelling percentage of 28.6 and kernel quality conforms to export grade W 210. Colour of apple is pink.

f) Ullal-3

It is a selection from 5/37 Manjeri and released in 1993. It is early in flowering (November - January) and fruiting period is very short (50-60 days). The fruiting is from January to March and sometimes starts from last week of December. It is a high yielding variety with average yield of 14.7 kg/tree. The nut size is medium with nut weight of 7g. The shelling percentage is 30 per cent and the kernel grade conforming to W 210 grade. The colour of apple is red.

g) Vengurla-4

The hybrid Vengulala-4 (Midnapur Red x Vetore 56) released in 1982 by Regional Fruit Research Station, Vengurala under Konkan Krishi Vidyapeeth, Dapoli. This is a popular variety with red colour apple. The variety is recommended for Maharashtra, Goa and coastal Karnataka. The average nut weight is 7.6 g, average shelling percentage is 31 and kernel is export grade (W 210). This is a cluster bearing variety and average yield is 19.08 kg/tree. The variety is becoming popular in other parts of the country also.

h) Vengurla-7

Hybrid 255 evolved at Regional Fruit Research Station, Vengurla under Konkan Krishi Vidyapeeth, Dapoli was recommended for release under the name "Vengurla-7" in the XIII Biennial Workshop of AICRP on Cashew held in November 1997. Vengurla-7 is a hybrid developed from the cross Vengurla-3 x M

10/4 (VRI-1). The percentage of perfect flowers is very high (40 per cent). Average yield is 18.5 kg/tree. It is a bold nut type with nut weight of 10g and shelling percentage of 30.5. Kernel count is W 180. The colour of apple is yellow with apple weight of 60g and with juice content of 75 per cent. Average weight of kernel is 2.9g. This variety is recommended for the Konkan region of Maharashtra and adjoining cashew growing regions of Goa and Karnataka.

i) VRI-3 (M 26/2)

This is a selection from seedling progeny of a high yielding tree collected from a village Edayanchavadi in South Arcot District of Tamil Nadu and was released in 1991. It has 12.1 per cent perfect flowers. The average yield of this variety is about 10 kg/tree, thus the increase over VRI-2 and VRI-1 being 35 to 39 per cent respectively. The nut size is medium with 7.2g nut weight and shelling percentage of 29.1 per cent. The kernel grade conforms to W 210 export grade. This variety is picking up fast among farmers of not only of Tamil Nadu but also of other states.

4.0 Nature and Causes of Postharvest Losses

In general, preharvest loss of raw cashewnuts is very minimal and mostly happens due to improper handling of the nuts while transporting from field to processing site. Hard outer shell withstands impact load preventing physical damage and corrosive oil which is present within the shell structure known as cashewnut shell liquid prevents microbial entry and protect against deterioration. But moisture plays vital role to enhance shelf life of the harvested nuts and hence nuts are dried under sun to reduce its moisture to safer level and stored in gunny bags of 80 kg capacity. Nuts, at times gets spoiled due to high pressure when stacked one over the other and change in the environment, but it is estimated that it will be within 1 per cent only. As far as cashew kernels are concerned, it is handled properly being a costly commodity and only low grade kernels *viz.*, LW, SWP or BB will mix with testa during peeling and will be lost. Cashew apple produced is underutilized primarily due to astringency and lesser shelf life. Around 10-15 per cent of the total cashew apple produced in the country is utilized for the preparation of alcoholic beverage called 'Feni' in the Goa state. Although many cashew apple products are developed, it could not attract consumers.

5.0 Postharvest Management

5.1 Maturity Standards

Cashew fruit has two parts namely cashew nut (true fruit) and cashew apple (pseudo fruit). Developments of these are not parallel as the cashew nut develops first and the pseudo fruit develops later. The whole process of development of both nuts and apples will be completed within 60 days from anthesis. Cashewnut develop still 30 days from anthesis and it gets hardened during subsequent 10 days. Afterwards the development of cashewnut ceases and cashew apple starts developing only after the cessation of growth of cashewnut and loses moisture during subsequent 10 days.

Raw cashewnut has pericarp which is hard in nature with spongy mesocarp containing caustic oil commercially known as Cashewnut shell liquid (CNSL). The edible kernel is wrapped up by leathery testa, is the embryo comprised of two large cotyledons, a radical and a plumule. The nut and apple develop in three well defined stages and are delineated by the colour of the nut which generally begins as pink after fertilization, turns green and finally grey at maturity. The embryo and pericarp grows to maximum size when the nut is green. The nut is considered as mature when starts turning into gray. The colour of the nuts changes from brownish green at fruit set to light green at ¼th growth of apple and thereafter turns grey irrespective of varieties. In the third stage of development, the growth rate of peduncle continues to increase. The size of the pericarp decreases slightly with the loss of moisture and the embyo remains constant in size. The total weight of the nut is high in the beginning of the third stage.

Development of cashew apple and nut has been studied by number of researchers in different cashew varieties. According to report, specific gravity could be considered as physical index of maturity of nut in cashew varieties. According to a report from Thailand the nut reached maximum size, approximately four weeks after fruit set, hardened in the subsequent week and declined in size by 10 per cent at harvest.

The growth of nut is faster during early stage of development than that of fruit. Changes in the composition of cashew nut and apple during development have been investigated by various workers. The amino acid content in kernel decrease with increased maturity. Cashew kernel lipid, free amino acids and CNSL have been shown to increase with increased maturity. Changes in the kernel sugars and starch were not uniform.

5.2 Harvesting

Raw cashewnuts are harvested in four different ways *viz.*, (i) collection after natural drop; (ii) thrashing by stick and collection; (iii) climbing on the tree and thrashing and (iv) shaking the tree. Studies indicated that about 72 per cent of the farms follow harvesting the nuts by thrashing with stick and collection method. Ultimately nut harvest by this method increased the quantum of immature nuts supplied for processing. Only 16 per cent of farms follow the collection of nuts after natural drop. Besides climbing on the tree and thrashing and shaking the tree to harvest the nuts amounts to 10 per cent and 2 per cent respectively. These two methods obviously increase the immature nuts and leads to inferior quality in the end product. Harvesting by thrashing or climbing also results in dropping of young flowers and ultimately reduces the production. Besides, the nuts have to be harvested at right stage. Normally fallen nuts are collected from the ground, but cultivators have tendency of plucking the nuts before the pseudo fruit develops completely in view of pilferage. Fully mature apples are plucked from the tree with the help of a long pole with a *sickle* tied to it at one end. The best method of harvesting the nuts is to wait till it falls, but this damages the pseudo fruit. If one intends to utilize cashew apple, it has to be plucked to avoid damage to the fruit.

5.3 Drying and Storage

Figure 1: Sun Drying of Raw Cashewnuts

Once the cashew fruits are either harvested or collected from the field, the nuts are to be separated from the cashew apple and sun dried for 2 to 3 days (Figure 1). Raw nuts at the time of harvest will have moisture content ranging between 16 to 25 per cent d.b. Sun dried nuts expected to have a moisture content of 5 to 6 per cent d.b. If the nuts are not dried and stored, it leads to fungal spoilage, resulting in poor quality of kernels. Fungi such as *Gonatobotryum*, *Helminthosporium* sp., *Corynespora* sp., *Alternaria* sp., *Verticillium* sp. and several species of *Aspergillus* have been isolated from stored cashew nuts.

Maximum permissible moisture content of raw cashew nuts is 8.7 to 9.1 per cent. It has been reported that moisture content of whole nut and kernel could be predicted from that of shell with more than 99 per cent accuracy. A higher kernel-shell weight ratio has been shown to be inversely proportional to moisture content. It has been reported that nut stored at 27 °C and at a relative humidity (RH) of 70 per cent had a moisture content of 9.2 per cent d.b. Nuts exposed to higher RH (>75 per cent) lead to mould infection within few weeks. Under high RH, nuts pick up moisture from the atmosphere till they attain equilibrium.

Gunny bag storage of 80 kg capacity is predominant in India and it has been shown that storage of raw nuts with a moisture content of 5 to 6 per cent for a period of 12 months at ambient temperature (25 to 30 °C) did not affect either processing or biochemical quality. Kernel rejects of 2 to 3 per cent obtained during processing of cashew by steam method has been shown to be influenced by immature nuts and floaters. Recent studies on effect of maturity on the processing quality have clearly indicated that shelling percentage, peeling outturn, whole kernel recovery increases while kernel rejects decreases with maturity. Application of insecticides or pesticides is prohibited to improve shelf life of raw cashewnuts in warehouses.

Raw cashew nut is kidney shaped one with 3.5 mm thick soft leathery outer skin (Epicarp) and thin hard inner skin (Endocarp). Cashew nut shell liquid (CNSL) is present in honeycomb structure located between these two walls of the shell. Kernel is inside the shell wrapped in a thin brown skin known as the testa (Figure 2).

5.4 Physical Properties

The size distribution of cashew nut is given in Table 3. In the sample, about 52 per cent of the nuts had length ranging from 30 to 35 mm termed the medium fraction, whereas 11.5 per cent and 36.5 per cent had length greater than 35 mm (larger fraction) and less than 30 mm (small fraction) respectively. The mean values of 100 measurements at moisture content 8.46 per cent db for the length, width, thickness

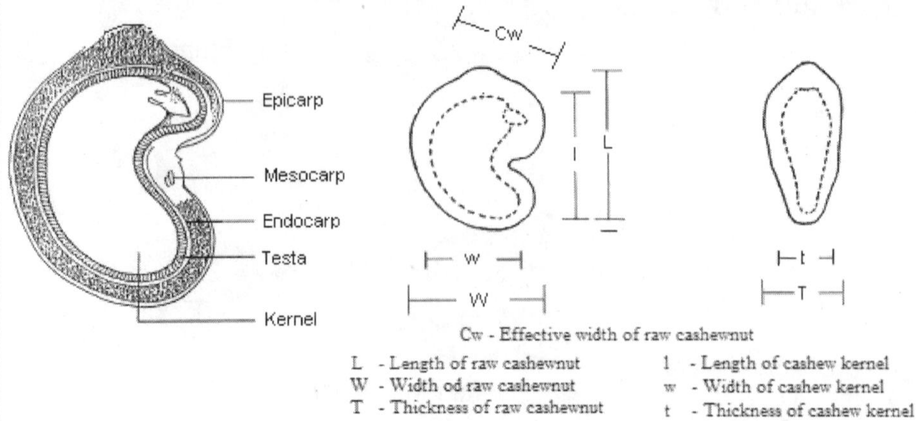

Figure 2: Geometry and Characteristics Dimension of Cashewnut

Table 3: Size Distribution of Cashew Nuts and Kernels at Moisture Content 8.46 per cent db

Particulars	*Ungraded*	*Size Category*		
		Large	*Medium*	*Small*
Length of Nut	23.06–40.06	> 35mm	30-35mm	< 30mm
Percentage of sample				
By number	200	23 (11.5)	104 (52.0)	73 (36.5)
By mass	1193	162.80 (13.7)	668.86 (56.1)	361.34 (30.3)
Average dimensions				
NUT				
(a) Length (L) mm		36.889 ± 1.655	31.868 ± 1.132	27.971± 1.454
(b) Width (W) mm		24.681 ± 1.629	23.270 ± 1.597	21.726± 2.068
(c) Thickness (T) mm		17.623 ± 2.046	17.297 ± 1.638	16.157± 1.655
(d) Mass (M) g		7.209 ± 2.245	6.447 ± 1.001	4.904 ± 0.842
KERNEL				
(a) Length (l) mm		28.169 ± 1.585	25.424 ± 1.473	22.548 ± 1.438
(b) Width (w) mm		12.491 ± 1.830	12.873 ± 2.662	13.300 ± 2.832
(c) Thickness (t) mm		12.556 ± 1.893	12.301 ± 1.350	11.578 ± 1.330
(d) Mass (m) g		2.353 ± 0.375	1.994 ± 0.356	1.611 ± 0.296
SHELL				
(a) Shell thickness		3.810 ± 1.117	4.041 ± 1.134	3.733 ± 0.423

and unit mass of the raw cashew nut were found to be 31.00 mm, 22.86 mm, 16.91mm and 6.04g respectively. The corresponding values of the kernel were 24.67 mm, 12.06 mm and 1.89 g respectively. A higher mass ratio of 3.20 of nut to kernel indicates relatively lower yield of kernel per unit weight of raw cashew nut. Shell thickness followed a linear relationship with moisture gain. The absorbed moisture, only after saturating the inter space available in honeycomb structure of shells moved to the inner skin and kernel area and is the reason for increase in its volume at once after soaking.

Bulk density decreased with increase in moisture content, shows that the increase in mass owing to moisture gain in the sample is lower than the accompanying volumetric expansion of the bulk. The true density of nut found to increase linearly from 1201 to 1240 kg m^{-3} with increase in moisture content of 3.15 per cent db to 20.06 per cent db. This may be due to higher weight increase of nut in comparison to its volume expansion on moisture gain.

The storage life of raw nuts in the godown also depends on the quality of nut and its environmental condition during storage. Fluctuating temperature is delirious during storage due to condensation of moisture in the godowns at certain pockets and consequent microbial growth. Atmospheric relative humidity (RH) plays a key role in safe storage of raw nuts as the nuts absorb or desorbs moisture to maintain equilibrium with the atmospheric relative humidity. Equilibrium moisture content of raw cashew nut was determined by static method at constant temperature (30 °C) and different relative humidity and presented in Table 4. The increase in the moisture content with the relative humidity exhibits desorption isotherm up to 74.12 per cent, but at 81.33 per cent it followed adsorption isotherm. Mould growth and deterioration took place after 28 days at 81.33 per cent after the commencement of experiment. The data were fit in the Henderson's equation and by solving simultaneous equations the values of the constants were found to be 7.09×10^{-4} and 0.865 for c and n respectively.

$$1 - R.H. = 7.09 \times 10^{-4} \times T \times M$$

Where R.H is the relative humidity expressed in decimal, T is the absolute temperature in °K and Me is the equilibrium moisture content in per cent db.

Table 4: Equilibrium Moisture Content of Raw Cashew Nuts at Various Relative Humidities

Saturated Salt Solutions	*Relative Humidity (per cent)*	*EMC (Per cent db)*
Potassium acetate	26.83	1.669
Potassium carbonate	43.91	3.371
Sodium nitrite	61.17	5.167
Sodium chloride	74.17	7.148
Potassium chromate	81.33	10.786*

* Mild fungus attack started after 28 days of commencement of experiment.

5.5 Quality Assessment

Grade standards have been developed for cashew nut, based on composite index. It has been found that the number of nuts per kg, moisture percentage and cutting test reduce the prices and these factors could explain 98 per cent variations in price. Among the quality variables considered, only three variables *viz.*, numbers of nuts, unit weight, floating test and nut length have been observed significant in price determination. These results could be used as a guide by the farmers for fetching better price in the market and by the buyers for getting product of ascertained quality.

5.6 Grading

Kernels are graded on the basis of specification prescribed by Govt. of India under the export (quality control and inspection) Act 1963, which recognizes 23 different export grades of kernels. The kernels are conditioned before packing in sealed tins. If the kernels are too dried at the time of packing, they are liable to breakage during transport by land and sea. If the moisture exceeds limit of 5 per cent kernels become susceptible to microbial and oxidative spoilage. Kernel grades of various cashew varieties released from cashew research centre in India is given in Table 5.

Table 5: Kernel Grades in different Cashew Varieties

Grade	*Varieties*
W-180	Priyanka, Vengurla -7, Ullal-4 and UN-50
W-210	NDR-2/1, Dhana, Vengurla -6. VRI-3. Selection-I and Selection-2
W-240	Ullal-I and Ullal-3
W-240	Chintamani -1
W-280	BLA-139/4, Kanaka, Vengurla -3 and Vengurla -4 and Ullal-2
W-320	Jhargram-I, BLA-39/1,K22, Bhubaneswar-1. Vengurla -1, Vengurla – 2 and Vengurla -5, VRI-I and VRI-2

Normally cashew kernels are sorted out manually based on size, surface colour and breakage (Table 6) and the average capacity of grading ranges from 80 to 120 kg and 35 to 60 kg for whole and broken kernels respectively depending on the degree of the skill. High degree of cleanliness is maintained in this section and workers are provided with apron and hair net. Minimum quality requirement are specified as processed kernels should be free from any deterioration; clean, practically free from any visible foreign matter; free from mould or rancidity; free from adhering testa and shell liquid and free of any foreign smell or taste. The moisture content of cashew kernels at the time of packaging is prescribed as 5 per cent d.b.

Table 6: Primary Criteria in Grading Cashew Kernels

Extent of breakage	Wholes (No breakage of kernel)
	Butts (2/3rd of kernel portion chipped off)
	Splits (Whole kernel separated into cotyledons)
	Pieces (All other broken)
Colour	White
	Scorched (Slightly brownish)
	Dessert (Deeply scorched with speckles)
Size	Wholes are graded according to size expressed as count per pound weight
	Pieces are graded according to the sieve size

5.7 Packaging

Cashew nuts are subjected to rancidity and become stale quickly due to hygroscopic nature and oxidation. Therefore packing needs to have low permeability of moisture and oxygen. Method of packing involves either vacuum or inert gas inside packing. At present the bulk of cashew nuts are packed by following two different techniques *viz.*, vita packing and molded vacuum packaging (MVP).

a) Mould Vacuum Packaging and Comparison

Introduction of alternative method of flexible packaging (Mould vacuum packaging) with nitrogen as inert gas is gradually gaining importance. After securing ISO 9000 and introducing HACCP (Hazard analysis and critical control points), importing countries are following stringent quality standards for cashew kernels and the present system of tin packaging is bulky and difficult to dispose of these tin containers. MVP system produces consistent rectangular blocks ranging in size from 500 g to 25 Kg. This gives a large and obvious improvement in the quality production with benefits of minimum movement in transit, handling, display, stock count etc.

b) Vita Packing

The graded kernels are packed in 11.34 kg capacity tins on weight basis which are subsequently evacuated and infused with carbon dioxide gas with the help of packing unit called "vita pack" to suppress the possible insect attack. The packed tins are then labeled as per the grades across the lid using special tamper proof adhesive. Two tins each containing 25 lbs cashew kernels are packed in corrugated carton box, which are bound by nylon strapping for export. The standard markings including brief description, name of packer, gross and net weight etc. are printed on the carton.

6.0 Domestic Marketing and Export

6.1 Import and Export Scenario

Over a period of time, a phenomenal growth is evidenced in the total quantity of raw cashewnuts processed in India. The pioneering spirit of the early entrepreneurs and the unsurpassed skills of Indian women in extracting whole kernels from the nuts were the major factors influencing birth and growth of cashew industry in India and for her preeminence as the world's largest processor and exporter of cashew kernel. A record quantity of 15.01 lakh MT of raw cashewnuts were processed during the year 2011-12, equivalent to 36 per cent of the global production. Although, a steady growth in the quantum of raw cashewnuts processed in this country, but the contribution of domestic nuts in the processing declined from 83.2 per cent during the year 1990-91 to 57.0 per cent during the year 2011-12 (Saroj and Balasubramanian, 2013). This clearly indicates that the sustainable production is not attained to operate the industry round the year to make it economically viable. Thus, there is a dependency of Indian cashew processors on the import of raw cashewnuts from African country.

Cashew kernels prepared from India is highly preferred by the importing countries and therefore, it stands in the first position as far as export of cashew kernel is concerned. Around 1.31 lakh MT of cashew kernels were exported during 2011-12 earning a foreign exchange to the tune of 4390 crores. The major markets for CNSL are the United States, United Kingdom, Japan and South Korea. Total world supply of CNSL is estimated at about 0.45 lakh MT with an average price of about US$ 300 per MT.

In India, only a small fraction of this potential CNSL is extracted from cashew shell obtained from industries following steam boiling mode of processing using expeller technique. However, these processors could change their processing methods provided the price of CNSL increases significantly.

6.2 Kernel Market

India was originally the dominant supplier of cashew kernel to the world market, but in the last 15 years, Brazil has also become a major exporter especially to the U.S. market. Tanzania and Mozambique were previously important suppliers but are now relatively small. In recent years, Vietnam has become a major source of kernel, as the government has assisted the development of the local processing industry by imposition taxes on the export of raw nuts. As far as the world market is concerned, there are two distinguished kernel markets *i.e.* International market, comprising 99.5 percent of world trade of cashew kernels which allows the use of inorganic material during production. Another market is the organic market, (0.5 percent world trade), relatively new development currently offers a 50 per cent price premium. The requirements in this industry are official certification that no chemicals or other banned substances are used during production and processing.

Major kernel markets in the world trade are i) United States Market, New York which is the world's largest cashew market; ii) European Market which includes Holland, Germany, UK and Rotterdam and iii) Australian Market, importing virtually 100 percent of its cashew requirements as domestic production is still negligible.

In previous years the dominant source of supply was India, however in the past few years Vietnam has become the major supplier because the trader believes Vietnamese crop is better graded and shipping is quicker and cheaper. Although major portion of the processed kernels in India were exported, consumption of cashew kernels at domestic level increased rapidly after 2000 A D, encouraging the cashew processors in India (Table 7). About 57 per cent of the totally processed cashew kernel is consumed by the Indians during the year 2010-11 indicating more or less 100 per cent in internal consumption over a period of 10 years. Strong processing potential with high demand of cashew kernels within India confirms sustainability cashewnut processing in India.

6.3 Raw Cashewnut Price

Price information is essential as it provides cashew business to determine the trade operations economically viable either immediately or in a long term. Variation in price of domestic and imported raw cashew nut are primarily due to the seasonality of indigenous nuts, increasing global demand of kernels and higher share of raw nuts in the processing cost.

The raw nut marketing in Kerala and Karnataka indicated that the nuts are sold immediately after harvest primarily due to the large processing facility in these states.

Table 7: Domestic Consumption of Cashew Kernels Processed from India
Quantity in lakh MT

Year	*Raw Nuts Processed*	*Expected Kernel*	*Kernel Export*	*Domestic Consumption*	*Per cent*
2000-01	7.21	1.51	0.92	0.59	39.07
2001-02	6.99	1.47	0.89	0.58	39.46
2002-03	8.27	1.74	0.98	0.76	43.68
2003-04	9.01	1.89	1.04	0.85	44.97
2004-05	9.88	2.07	1.01	1.06	51.21
2005-06	11.23	2.36	1.27	1.09	46.19
2006-07	11.38	2.39	1.14	1.25	52.30
2007-08	12.13	2.55	1.19	1.36	53.33
2008-09	13.01	2.73	1.09	1.64	60.10
2009-10	13.65	2.87	1.08	1.79	62.32
2010-11	11.68	2.45	1.06	1.39	56.78

Source: Directorate of Cashew and Cocoa Development, 2011.

In Goa, nuts are accumulated over a period of time and brought to the market during a particular period in a month. Besides, Agricultural Produce Marketing Centre (APMC) regulates the marketing of raw cashews in the yard and brings it to the marketing stream (Balasubramanian, 2003). Large growers in Andhra Pradesh tend to hold the nut and sell it in large volume. The factors influencing price of domestic raw nuts are i) Quantity of domestic production; ii) Kernel price in the international market; iii) Quantity of kernel exported; iv) Quantity of raw nut imported and v) Imported raw nut price.

Wide gap between the domestic production and the demand of the finished goods at premium price and to utilize the labour engaged in processing in efficiently throughout year led to import of raw nut in to India. Processors execute forward trading with international kernel market in the beginning of the year and forced to import raw nut during season to fulfil the contract. Monthly imports are fairly even, but may decline in May and June when the factories are occupied with the local crop. The different harvesting period of cashew in the world ensures the supply of nuts around the year. This helps to have control on imported nut price and the Indian processors are assisted by not carrying long-term stock. The nut supply from east African countries is improving due to closure of the processing facility in that country (Balsubramanian[2], 2002). Although, India has strong domestic market for the kernels produced, major advantages of exporting kernels over domestic trade are I) export promotion in terms of incentives *i.e.* 7 per cent of value of total exported kernel would be imported without tariff), ii) rapidly increasing equivalent currency rate and iii) kernel demand in bulk quantity.

7.0 Conclusion

Cashew has major economic, social and environmental benefits, not only for rural populations but also for the countries as a whole. In order to maintain India's position in the global market, certain research and development strategies have to be planned carefully and need to be implemented in time. Common quality standards for raw cashewnuts and cashew kernels need to be drafted and after assessment at various levels brought in to practice to benefit all stake holders in the cashew value chain. Price stabilization mechanisms for cashew need to be worked out to explore the possibilities of further motivating the farming community. A major challenge to increasing the sales of low grade or broken kernels is to find industrial food processors likely to develop and commercialize cashew-based products or products using cashews as an ingredient on a large scale. Value addition to plain cashew kernels, improvement in packaging technique, branding of cashew, development and promotion of cashew byproducts *i.e.* cashew apple and cashewnut shell liquid needs to be given utmost important. National laboratories need to be strengthened and accredited for certification in order to develop confidence and assure quality at consumer's level.

8.0 Important References

Acland, J.D. 1975, East African crops. FAO and Longman, 5: 29–32.

Antarkar, M.V and Joshi, G.D. 1987, Studies on physical indices of maturity of some cashew nut varieties. *The Cashew,* 1:18-21.

Augustin, A and Unnithan, V. K.G. 1981, An attempt on maturities of cashew apple. *Indian Cashew J.,* 14:9-11.

Balasubramanian, D and Joycy, R.L.K. 2012, Performance Evaluation of Mechanized Shelling Machine for steam treated raw cashewnuts. In: Abstract of PLACROSYM XX. Mechanization of sustainable productivity. Eds. Radhakrishnan, B, Rajkumar, R and Premkumar, R. Coimbatore, Tamilnadu, India, P:127.

Balasubramanian, D. 1997, Equilibrium moisture content of raw cashew nuts. *Cashew Bulletin,* 34(2): 2-3.

Balsubramanian[2], D. 2002, Speculation of raw cashew nut Price – An over view. In: Proceedings of PLACROSYM XV. Sustainability of Plantation Crops through Integrated Approaches for Crop Production and Product Diversification, Mysore, Karnataka, India, P: 205

Balasubramanian[2], D. 2000, Status report of cashew industries in Dakshina Kannada. *The Cashew,* 14 (2): 12-25.

Balasubramanian[3], D. 2000, Physical properties of raw cashew nuts. *J. of Agricultural Engineering Research.* 78(3): 291-297.

Balasubramanian, D. 2010, Design and development of radial arm type cashew kernel extracting machine" *Agricultural Mechanization in Asia, Africa and Latin America,* 42(2): 49-55.

Balasubramanian, D. 2012, Comparative performance of mechanized peeling machines for unpeeled cashew kernels – A case study. In: Abstract of PLACROSYM XX. Mechanization of sustainable productivity. Eds. Radhakrishnan, B, Rajkumar, R and Premkumar,R. Coimbatore, Tamilnadu, India, P:130.

Hariharan, M and Unnikrishnan, K. 1985. Effect of gibberellic acid and total protein contents in developing kernel of cashew. *Acta Hort*, 108:25-30.

Joseph, K.T. 1981, Seed-borne fungi on cashew (*Anacardium occidentale* L). *Cashew Causerie*, 3: 12.

Mathew, A.G. 1995, Improved shelling machine. *Cashew Bulletin*, 32 (4):

Mohankumar, G. N, Narayanaswamy, P and Mukashi, A. N. 1984, Fruit growth and development of cashew. *J. Plantation Crops*, 12: 81-85.

Nagaraja, K.V. 1998, Quality of cashew kernels in relation to export. *The Cashew*, 12: 143-148.

Okwelogu, T. N and Mackay, P.J. 1969, Cashew moisture relations. *J. Sci. Food Agric.*, 20:697-702.

Rao, C and Joshi, G.D.F. 1987, Studies on physical indices of maturityof some cashew nut varieties. *The Cashew*, 1(4): 180-21.

Saroj, P. L. and Balasubramanian, D. 2013, Cashew Industry in India –a sustainable road map. *Indian Horticulture*, 58(1): 9-15.

Sumrit Feungehan, Taweekiat Yimaswat, Sorot Chindaprasert 1989. Physiological development of cashewnut on canopy flower and fruit. *Khon Kaen Agric J*, 17:239-253.

Thivavarnvongs, T.O. and Tand, K.O. 1995, Design and Development of Compact Sized Cashew Nut Shelling Machinery- Testing and Evaluation of Manual and Semi-Automatic Shellers. *Journal of the Japanese Society of Agricultural Machinery*, 57(2): 57-65.

Chapter 28
Chilli

Surya Kumari Surepeddi

Principal Scientist

Giridhar Kalidasu

Scientist

AICRP on Spices, Dr. Y.S.R. Horticultural University,
Horticultural Research Station, Lam, Guntur – 522 034, India

E-mail: surya.surepeddi@gmail.com; E-mail: gkalidasu@yahoo.com

1.0 Introduction

Chilli (*Capsicum annuum* L.) belongs to the genus *Capsicum*, many members of which are valued for their taste, colour and spice. They are widely appreciated for their active principles and essential oil. Chilli was consumed by the people of south India and Central America as early as 7000 B.C., but it was cultivated some 2000 years later. Many civilizations were known to use this species, especially the Mayans and the Aztecs. Though the crop was domesticated long ago, it is considered as a new crop since its use drastically increased only in the last few hundred years.

A major portion of the genus *Capsicum* arose in south central Bolivia with subsequent migration to the Andes and low land Amazonia accompanied by radiation and speciation. Five species *viz. C. annuum* var. *annuum*, *C. chinense*, *C. frutescens*, *C. baccatum* varieties *pendulum* and *umbilicatum*, and *C. pubescens*were domesticated bynative Americans. Of the five domesticated species, initially *C. baccatum* and *C. chinense* were cultivated in northern South America while it is probable that *C. annuum*, *C. frutescens*, and *C. pubescens* were domesticated in Mexico or northern Central America. Following the initial visits of Spanish to the Americas, the chilli was brought to Europe and from there to most tropical, subtropical, and temperate zones. Among the several species, the most economically important species in the world is *Capsicum annuum*. Pungent cultivars (*ajies, paprika, chilies, and hot peppers*) are generally used as spice; whereas the sweet types (*sweet pepper, bell pepper, pimento*) as vegetables. Chilli pepper generally refers to a pungent chilli variety. Considering the trade, consumption and volume of production, red pepper products, pungent and non-pungent, represent some of the most important spice commodities in the world. Dry chilli is extensively used as a spice in the entire globe and along with other spices and condiments it is used to prepare various spice mixtures and curry powders. It is also used for sauces, chutneys, pickles, frankfurters, sausages etc. Bird chilli is used in making hot sauces as pepper sauce, tabasco sauceand certainalcoholic beverages like Mandrani of West Indies.The hot and pungent fruits are employed in ethnic, traditional and modern medicine. Its pharmacological action as counter irritant, stomachic, gastro-protective, neuropathy, diabetic neuropathy, beneficial effect on metabolic and arterial function and carcino-protective effects are widely appreciated. Capsaicin has been used in different allopathic creams to alleviate pains as it interferes with the sensory nerve terminals. Capsaicin incorporated creams are widely used in dermatologic therapy and proven useful in preventing chronic pain associated with post-herpetic neuralgia, diabetic neuropathy, and other pain syndromes. Pepper spray is also used in defence and self defence applications to ward off enemies or attackers. High heat oleoresins are used for this purpose. Apart from its noted medicinal values, chilli recently attained status of beverage component with its capsaicin being used an active ingredient because of its pleasurable and euphoriant effects after ingestion.

2.0 Status of Cultivation

The prominent world producers of chilli are India, China, Pakistan, Korea, Mexico and Bangladesh.

India is the largest producer, consumer and exporter of dry chilli in the world. Among the spices produced in the country, chilli (with 22 per cent production share of all the spices) is the most important one followed by garlic, turmeric, ginger and seed spices. India contributes about 39 per cent to the total world production. It commands a lion's share of exports with 49 per cent of global trade by volume. The crop is cultivated in all the states and Union territories of the country and covers 7,94,000 ha area with a production of 13,04,000 metric tonnes. Important chilli growing areas in India are listed in Table 1.

The production of chilli in India is dominated by Andhra Pradesh which contributes nearly 53 per cent to the total production. Karnataka and West Bengal are the second largest producers contributing 8 per cent each to the total production followed by Madhya Pradesh (7 per cent), Odisha (5 per cent), Gujarat (5 per cent), Maharashtra (4 per cent),Tamil Nadu (2 per cent) and others about 8 per cent during 2012-13 (Figure 1). The major chilli growing districts in Andhra Pradesh are Guntur, Warangal, Khammam, Krishna and Prakasam. Guntur is the biggest chilli producing region, contributing 30 per cent to the total production of Andhra Pradesh. Chilli production in India is moving northwards because of increasing demand from diversified sectors and changing consumption patterns. Dry chilli production has

Table 1: Important Chilli Growing Areas in India

State	*Important Chilli Growing Areas*
Andhra Pradesh	Guntur, Khammam, Warangal, Prakasam, Kurnool, Cuddapah, Adilabad, Medak, Nizamabad
Assam	Nalbari, Darrang, Dhubri, Nagaon, Sonitpur
Gujarat	Banskantha, Jamnagar, Junagarh, Rajkot
Karnataka	Belgaum, Dharwad, Gadag, Bijapur, Davangere, Raichur, Bangalore, Haveri
Maharashtra	Nasik, Pune
Manipur	Imphal, Churchandpur, Thoubal
Meghalaya	West Garo Hills, South Garo Hills, North Garo Hills
Nagaland	Kohima, Dimapur, Phek, Wokha, Mokokchang, Mon Zuenhoboto
Tamil Nadu	Coimbatore, Salem, Dindigul, T'hruvallar, Thiruchapalli, Ramanthpuram, Sivanagar, Karur, Namakkal, Dharampuri, Virudhnagar

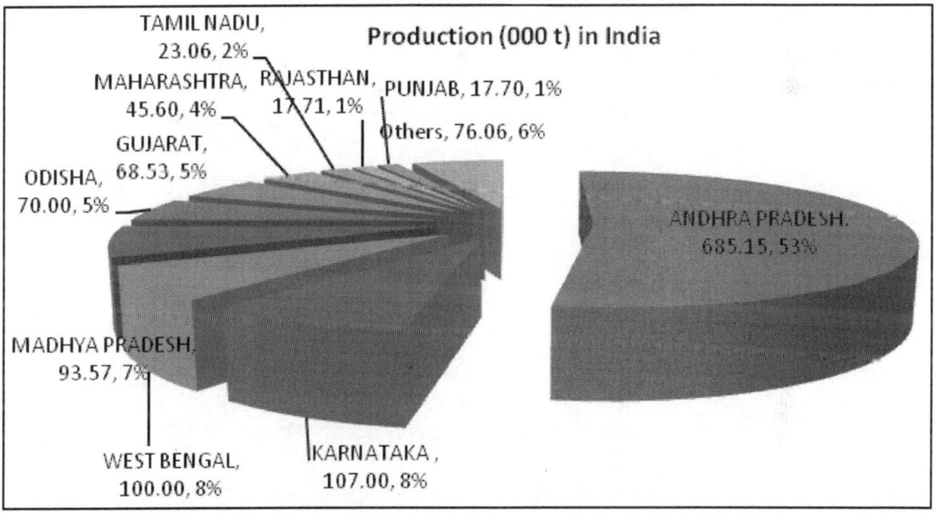

Figure 1: Dry Chilliy Production and Production of Share of different States in India, 2012

risen by nearly 156 per cent from 5.091 lakh tonnes in 1980 to 13.04 lakh tonnes in 2012 - 13. Rising export demand coupled with higher price realization in the domestic market have motivated farmers to bring more area under chilli cultivation. The average productivity of the crop is reported to be the highest in Uttarakhand (3600 kg ha^{-1}) followed by Andhra Pradesh (3262kg ha^{-1}) in 2012.

3.0 Commercial Varieties

3.1 Globally Traded Cultivars

In the genus *Capsicum*, enormous morphological variability exists for fruit colour, size, shape, pungency, flavour and aroma. Based on fruit size, shape and degree of pungency, a large number of horticultural types are recognized worldwide and at least 20 types are cultivated on a large scale in one or other parts of the world. Some of these fruit types such as *ancho, bell, jalapeno, pasilla, new mexican, yellow wax* have specific traits for fresh use and flavour pungency and for processing. In India, a number of genotypes with specific fruit size, shape and quality attributes are commercially cultivated in different regions. Codex standards prescribe pungency expressed in Scoville Heat Units (SHU) in to four categories *i.e.* mild (900-1999 SHU), medium (2000-19999 SHU), hot (20000-100000) and extra hot (> 100000 SHU). Due to location specific consumer needs and nature of end use; colour, shape, size, pungency and flavour vary widely according to local preference. The fruit quality and type depends mostly on the target market. Considering the fresh market,low pungent pimentos are preferred in parts of Europe.In India extremely hot Jwala and hot types are preferred. Similarly, in dry chilli for best quality hot to mild paprika preparation, half a dozen types varying from long red wrinkled to round red cherry peppers are commonly used.The crop is generally distinguished by its pod type and several hundred pod types are grown worldwide. The important chilli varieties of the globe in different markets are presented in Table 2 and the commercial uses of chillies are presented in Table 3. Numerous subtle flavours of chillies exist and from experience one can know these flavours.For example *ancho*is sweetish, *mulatto* is chocolaty, *mirasol* is fruity, and *chipotle* is smokey.Grinding the pods produce one

Table 2: Important Chilli Types and Varieties

Species	*Type*	*Important varieties*
C. annuum L.	Ancho or Mulato or Pablo Type (wide throated)	Ancho 101
	Asian types	Pusa Jwala and many Indian varieties, Takanotsume, Shishito, Santaka, Prik Kee Nu, Prik Chee Faa and Thai Chile.
	Bell types (blocky fruits)	California Wonder, Dove, Golden Bell, Valencia, Lilac Belle and Mexi-Bell.
	Cascabel types (rattling pod, ball chilli)	Cascabel Local (Mexican)
	Cayenne types	Large Red Thick, NuMex varieties
	Cherry types (cherry pods)	Chilhuacles, Beslers Cherry, Hot Cherry Pepper, Cherry Pick and Cherry Bomb.
	Chiltepin (small round pods) or Chile piquin (small elongated pods) types	NuMex Bailey Piquin
	Chipotle types (smoked red ripe jalapeño)	Jalapenos
	Costeño types (used for sauses)	Costeño and Costeño Amarillo
	De Árbol types (treelike plant, mature plant resembles a small tree)	De Árbol Purple

Contd...

Table 2–*Contd...*

Species	*Type*	*Important varieties*
	Ristra types (fruit wreaths)	NuMex Sunburst, NuMex Sunflareand NuMex Sunglo
	Jalapeño types (named after town of Jalapa, Mexico; thick walled conical fruits)	Early Jalapeno, NuMex Jalmundo, NuMex Piñata, NuMex Primavera, and NuMexVaquero.
	Mirasol types (looking at the sun, upright fruits)	NuMex Mirasol, Guajílloand Pulla.
	New Mexican types (long green pods)	ChimayóEspañola, ImprovedNew Mexico No. 6-4, NuMex Heritage 6-4, NuMex Big Jim and NuMex Heritage Big Jim, NuMex Conquistador, NuMex R, NakyNuMex, SweetNuMex Sunrise, NuMex Eclipse, and NuMex Sunset, NuMex Joe E. Parker and Sandia.
	Onza types (short slender fruits)	Onza Roja and Onza Amarillo
	Ornamental types	Black CubanChilli, ChilliJigsaw, NuMex Centennial, NuMex Twilight, NuMex Christmas, Peter Pepper and Poinsettia.
	Pasilla types (little raisin types; dark brown oolour and raisin like aroma)	Apaseo, Salvatierra and Holy Mole.
	Piment D Espelette (heirloom chile pepper from Espelette in South wostern France)	Espelette
	Pimiento/Pimento (sweet heart shaped, thick-walled fruit)	Perfection Pimiento, Pimiento Solect, Pimiento Sweet and Pimiento L
	Serrano types (meaning from the highland or mountain cylindrical, pendant, and bluntly pointed pods)	Balin and Tam Hidalgo and Serrano Tampiqueño.
	Wax/güero types (yellow waxy immature fruits)	Floral Gem, Mississippi Sport, and Santa Fe Grande
C. chinense Jacq.	Fruits are generally wrinkled, slight to clearly segmented and very hot	Ají Limo, Ají Panca, Bhut Jolokia, Cajamarca, Charapita (wild), Chocolate Habanero, Datil, Fatalii, NuMex Suave Orange, NuMex Suave Red, Orange Habanero, Red Carribean, Rocotillo, and Scotch Bonnet.
C. frutescens Roxb.	Borne erect, extremely hot fruits, fruit flesh often soft	Tabasco, Greenleaf Tabasco, Malagueta, and Siling Labuyo.
C. pubescens Ruiz and Pav.	Fruit flesh thick, firm	Canario, Manzano, Peron and Rocoto Longo.
Capsicum baccatum L.	Pods with distinctive fruity flavour with low to high heat types	Ají amarillo, Ají ayucllo, Ají Limon, Christmas Bell, and Omincolour.

Source: Dewitt and Bosland, 2009.

flavour, toasting before grinding produce another, and soaking the chillies in water produce yet another.Many pod-types are central ingredients in ethnic dishes such as ancho and New Mexican inchile rellenos, pasilla inmole sauce, and serrano inpico de gallo.

The details of important chilli types and varieties across the globe are presented in Table 3.

Table 3: Commercial Uses of Chilli and Preferred Types

Capsicum Type	*Purpose*	*Preferred Types*	*Products*
All edible types (*Capsicum* spp.)	Fresh whole fruits	All types	–
C. annuum L.	Canned or bottled, frozen, pickled or brined, fermented fruits and sauces	Pimento, cherry peppers, jalapeño, serrano, yellow wax, pepperoncini, Thai chilli and New Mexican	Pimientos, Canned jalapeño, convenience food, pickled Pepperoncini, stuffed pepper, Ceviche and Sambal oelek
C. chinense Jacq.	Fermented sauces	Habanero, Scotch Bonnet, Yellow lantern chilli and King chilli	Puree, Salsa, Jerk dishes, Habanero Hot sauce, Scotch Bonnet Sweet Sauce
C. frutescens Roxb.	Fermented sauces	Tabasco pepper, Malagueta chilli and Thai pepper	Creole cooking and to make the famous hot Tabasco Sauce
C. annuum L., *C. frutescens* Roxb., *C. chinense* Jacq.	Spice extract, Paprika and chilli oleoresins	Hungarian paprikas, Pimento and Indian Sweet types	Oleoresin Capsicum (OC), Oleoresin Paprika (OP)
C. frutescens Roxb., *C. chinense* Jacq.	Pharmacological and defence (Capsaicin)	Red Savina, Habanero and Naga jalokia (Tejpur)	Oleoresin Capsicum (OC)
C. annuum L., *C. frutescens* Roxb., *C. chinense* Jacq.	Whole pods, Paprika powder	Many types - New Mexican, cayenne, ancho, pasilla, mirasol, piquin, de arbol, and Indian hot and sweet types	Consumer unit packed pods, crushed pods, chilli powder, and spice mixtures.
	Pharmacological (Capsanthin)	Hungarian paprikas, Pimento and Indian Sweet types	Oleoresin Paprika (OP)
	Decoration and pot plants	Christmas peppers	Colourful strings (ristras), potted peppers and Yellow lantern plants

3.2 Commercially Traded Types in India

The major types grown in India are pungent ones and a few are low pungent types. The most important types traded are sannam, mundu, wrinkled and bird's eye chillies. There are also many other varieties like Ellachipur Sannam, Hindpur-S7, Kanthari White, Madras Pari, Kashmir chilli, Nalchetti, Ramnadu Mundu, Sattur-S, Byadagi, Warangal Chapata and Tomato Chilli. Each type is known for its intrinsic value, characteristics and specific market. The colour and capsaicin content are presented in Table 4.

Table 4: Colour and Capsaicin Content of Major Indian Chilli Varieties

Chilli Type	*Location*	*ASTA Colour Value*	*Capsaicin (per cent)*	*Scoville Heat Units (SHU)*
Birds EyeChilli	Mizoram and somelocations of Manipur	41.70	0.589	88350
Byadagi (Kaddi)	Hubli, Dharwar district of Karnataka and Goa	156.90	Negligible	-
Ellachipur sannam	Amaravathi District of Maharashtra	70.40	0.200	30000
Guntur Sannam	Guntur, Warangaland Khammam, districts of Andhra Pradesh	32. 11	0.226	33900
Jwala	Kheda, Mehsana and in South Gujarat	-	0.400	60000
Kanthari-white	Kerala and some parts of Tamil Nadu	2.96	0.504	75600
Kashmir chilli	Himachal Pradesh, Jammu and Kashmir and also in sub-tropical regions of North India	54.10	0.325	48750
Hindpur	Hindpur in Andhra Pradesh	33.00	0.240	36000
Ramnad Mundu	Ramnad district of TamilNadu	32.95	0.166	24900
Tomato Chilli	Warangal, Khammam, East and West Godavari districts of Andhra Pradesh	125.26	0.170	25500

Source: Prasath and Ponnuswami, 2008.

4.0 Nature and Causes of Postharvest Losses

Chilli is unique among vegetables, because green form is used as a vegetable and dry form is used as a spice. Green chilli, like other vegetables is a perishable commodity. While the dry pods and ground chilli due to improper postharvest management loose the important quality traits like colour and pungency giving poor appeal to consumers and low returns to stake holders. Due to the perishable nature, quality of fresh fruits quickly changes during storage, transit and marketing. Spoilage due to wilting, shrivelling, pathogenic disorders, water loss, etc. after harvest is common under improper postharvest management. Postharvest losses in fresh chillies go up to 20 per cent in market chain and maximum loss occurs at wholesale trader level. The losses are mainly due to sorting and storage rather than damage in transit. Apart from physical changes like bruises and wrinkles, loss of fruit weight is another serious damage to fresh chilli which distances the consumer away from buying the vegetable. Loss of glossiness, green or red colour and lowering of ascorbic acid content is usually associated with physical changes that occur due to ageing. The fungal and bacterial disorders also cause severe loss of shelf life and quality. Among these most important are *Anthracnose* and bacterial soft rot which are quite common causing damaged particularly in storage. Chillies are very sensitive and easily get damage while handling. They also loose shelf life and quality quickly under improper storage conditions. Further, they are quickly damaged by too low or too high temperatures during storage. Utilization of proper harvest and postharvest handling methods are essential for maintaining acceptable quality and maximizing shelf life.

4.1 Pre-harvest Factors

Proper pre harvest crop management solves many postharvest problems. Poorly managed crop ultimately gives low quality produce. Crop nutrition plays important role in getting good quality produce. Maintaining good, long-term soil health with optimal supply of nutrients to plants inthe growing season will help in achieving qualityfruits throughout the season. Irrigation is one such factor that helps not only productivity of the crop but also desirable quality and shelf life of the produce. A fruit with optimum moisture content, tends to thrive long during storage, transit and marketing overcoming the moisture loss during these processes. Water stress during the growing season can affect the size of the fruit, and lead to soft or dehydrated fruit that is more prone to damage and decay during storage. Excess moisture in the harvested fresh chillies increases the incidence of postharvest diseases and development of aflatoxins. Harvesting during rainy periods encourage the postharvest decay on storage of fruits. Extended shelf life of fresh chillies produced under green house conditions was observed compared to the produce from open field conditions. Maintaining ideal soil moisture content not only increases the total fruit biomass on the plant but also reduces the incidence of blossom end rot. A series of experiments conducted on sweet pepper in Netherlands revealed that increasing EC levels of the nutrient solution reduced the shelf life, while it improved flavour, measured in sensory assessments. Further growing environment with 24 h optimum humidity of 80 per cent gave best flavour and shelf life.

4.1.1 Chilling Injury

Chilling injury includes fruit softening, water-soaked spots, surface pitting, lack of colour development, discolouration of the seed cavity, and increased susceptibility to decay. Chilling also causes browning and decay of the stem. Ripe or coloured peppers are less chilling sensitive than green peppers. Fresh chillies for export markets are usually stored at low temperatures between 7 to 12 °C. Chillies will develop surface pitting in a few days when stored at 2 °C. Fresh green chillies can be stored up to three weeks without much postharvest losses when maintained at 10 °C (R.H. 80-90 per cent). However, storing at temperatures lower than 5 °C (R.H. 80-90 per cent) promoted chilling injury in fruits. Fairly low levels of chilling injury can be maintained up to 28 days of storage when fruits are held at 10 °C. Symptoms of chilling damage usually do not appear until peppers are moved from cold storage to ambient temperature.

4.1.2 Ethylene Injury

Fresh chillies are very sensitive to the natural ripening hormone ethylene. The presence of even minute quantities can cause a loss of chlorophyll and discolouration of the green pigment in chillies, resulting in a mosaic of yellowish, chocolate, and red colours. The combination of ethylene with high temperature accelerates the rate of loss in green pigment. Ethylene producing crops such as bananas, plantains and tomatoes should not be stored or shipped with green coloured peppers. The chilli fruits particularly sweet peppers should never be exposed to ethylene.

4.2 Postharvest Diseases

Chillies are susceptible to a number of postharvest diseases, most of which need entry point through mechanical damage or weakening of the tissue before they can penetrate into the fruit.The waxy skin of chillies offers protection against pathogen infection, but it is easily damaged while handling. Bacterial and fungal pathogens may also enter the fruit through the severed stem tissue and natural openings around the calyx. The most common postharvest diseases are bacterial soft rot, grey mold, rhizopus rot, watery soft rot, black mold, anthracnose, and sour rot. The incidence of these diseases is always higher during wet, rainy weather. Proper pre harvest disease control practices along with careful handling, cleaning, and grading of the fruit, and appropriate postharvest temperature control will minimize the occurrence of disease. Dipping of fresh fruits in two percent solution of sodium hypochlorite and sodium chloride and subsequent storage at 10 °C with 90-95 percent humidity resulted in the lowest decay of fruits. Bicarbonate salts have broad spectrum antimicrobial properties and are Generally Recognised as Safe(GRAS) compounds. Dipping of harvested red peppers for 2 minutes in 1 to 2 per cent potassium bicarbonate reduced the decay to a commercially accepted level of 5 to 8 per cent where as higher concentrations reduced fruit quality significantly.

Methods to reduce posthavest decay are discussed below:

a) Anthracnose/Fruit Rot

This is an important postharvest disease of chilli and caused by the fungus *Colletotrichum capsici*. Field infection is the primary reason for the damage. Infection usually occurs, when fungal spores are splashed on to developing fruits. The spores usually remain in a dormant state until conditions are right for sporulation and growth of the fungus. Ripe fruits are more susceptible to this disease. Initially small, water soaked lesions appear on the fruit surface. These lesions rapidly grow and appear as sunken, black in colour. As the infection progresses, salmon-coloured spores appear either scattered or in concentric rings within the lesions. Field sanitation is the most important factor in the management of the disease. It is essential to remove infected crop debris and periodic applications of protective fungicides. Storing the fresh fruit at 7 °C after harvest will also minimize decay by the pathogen. In dry chilli preparation, the disease is favoured by high moisture content in chillies and slow drying process especially by traditional method of drying results in deterioration and loss of quality beside aflatoxin contamination. Drying chillies in solar tunnel dryer for more than 24 hours results in rapid removal of moisture and significantly low incidence of fruit rot.

b) Bacterial Soft Rot

It is the most destructive postharvest disease in chillies and is caused by soil-borne bacteria *Erwinia carotovora*. The field infection occurs during rainy weather when soil containing the bacterium is splashed onto the susceptible fruit. It is also carried when contaminated water is used in postharvest cleaning operations. The pathogen usually enters the fruit through any wound, crack, opening or any injury caused during insect feeding. Excess nitrogen fertilization is associated with increased susceptibility to soft rot. Initial symptoms begin as light coloured, water soaked areas around the edge of wounds on the fruit or stem. The water-soaked areas quickly enlarge, tissues around the infected site begin to soften and the fruit eventually turns into a mushy watery mass. Postharvest infections are particularly damaging, because if one fruit decays the disease may eventually spread to all fruit in the same container. Bacterial soft rot is easily recognized by a foul-smelling odour as the fruit decays. In later stages, bacterial ooze may develop from affected areas, and secondary organisms follow, often invading the rotten tissue. Both green and ripe fruit may be affected and the affected fruit hang from the plant like a water-filled bag. The optimal temperature for bacterial soft rot growth is 27 °C to 30 °C and the minimum temperature for growth is 3 °C. However, the pathogen can effectively damage even at 10 °C if the conditions are conducive for pathogen development. While field infestation can bemanaged by controlling insects that create wounds for infection the postharvest decay is reduced by picking fruit when it is sufficiently ripe and dry, avoiding injury during handling, and holding the fruit at 7 °C. If fruit is to be washed after harvest, the water should be properly chlorinated and the fruit should be dried as quickly as possible after washing. Pungent pepper cultivars usually have a defined abscission zone and do not suffer stem infections unless damaged during harvest.

c) Black Mold

The disease is caused by the fungus *Alternaria alternata*. The pathogen attacks fruits that are having injuries, but does not usually spread from fruit to fruit. Initially, small, circular, slightly sunken spots associated with cracks or other surface injuries are seen on the fruit. These spots enlarge into sharply sunken lesions, eventually becoming covered with a greyish to black mold. The lesions show alternating light and dark brown concentric zones typical of pathogen attack. The pathogen also attacks and grows on placenta and seeds. Storage of the fruit at low temperatures will significantly retard the development of black mold.

d) Grey Mould

It is the second most important postharvest disease of chillies after soft rot and is caused by the fungus *Botrytis cinerea*. Initial infection can be identified by small cream coloured specks on wounded areas of the fruit surface. These spots enlarge into circular water soaked lesions covered with a grey mold. The optimal temperatures for growth of the fungus are between 18 °C to 23 °C. The fungus may occasionally attack green chillies during cold storage. Proper field sanitation and prevention of wounds on the fruit helps in reducing the incidence of grey mold.

e) Rhizopus Rot

It is a common postharvest disease of chillies. The disease is caused by the fungus *Rhizopus stolonifer*. The pathogen primarily attacks wounded areas of the fruit tissue or enters through the broken stem. Small water-soaked lesions are seen initially which quickly enlarge without discolouration. In advanced stages entire fruit is attacked by the pathogen. Greyish white masses of mold develop over the wounded area, which eventually turn black. A clear liquid is released from diseased tissues. Nests of mold and decaying fruit may form within a carton of packed fruit. High temperature and humidity and fresh wounds promote disease development. The main ways to control *Rhizopus* rot are to sanitize all field containers and grading surfaces, prevent wounding of the fruit tissue, dry wet surfaces of the fruit and cool the fruit to 7 °C as soon as possible.

f) Scelorotium Watery Soft Rot

The disease is caused by the fungus *Sclerotinia sclerotiorum*. It is seen under high humid conditions during rainy season. Air borne spores of this soil borne pathogen, anchors in wounded areas of the fruit and establishes there. Small lesions develop on the fruit surface; later develop in to a white mold. The fruit parts affected becomes soft and watery. The pathogen spreads easily among the fresh fruits stored in store houses or cartons. The disease is controlled by using good pre harvest sanitation practices. The disease attack can be mitigated by storing the fresh chillies at 7 °C.

g) Sour Rot

This is a minor disease occasionally seen in fresh chilli fruits. It is caused by the soil-borne fungus *Geotrichum candidum*. Initial symptoms appear as pale greasy water-soaked lesions. The fungus enters through wounds at cracks, cuts, or skin punctures. The affected tissues remain fairly firm until the decay is quite advanced. A typical sour odour is given off by the fruit as it rots. The disease can be controlled by storing the fruit at 7 °C, preventing injury to the fruit during harvest and handling, and maintaining good ventilation during storage.

5.0 Postharvest Management

5.1 Maturity Standards

Size of the fruit is the most widely used harvest index in chilli. Apart from size, the calyx and stem should be fresh, firm and green. External colour is another widely used index of harvest maturity. Most chilli cultivars change their skin colour as they become fully mature. The colour of the fruits vary from light to dark green, pale green or whitish green finally turning to red in most of the cases. In general fruits are harvested according to consumer preference.

5.2 Harvesting

The crop takes about 40 to 50 days to flower after transplanting and a further 10 to 20 days to develop fruit suitable for green harvest. Harvesting is a crucial operation in chilli which in fact decides its postharvest quality, shelf life and marketability. Harvesting of chillies depends on the purpose for which it is grown. The fruit is non-climacteric in nature;and stagered harvesting is the norm as the flowering is asynchronous. The effect of harvesting at different maturity stages of the fruit indicated that the colour values were generally highest at the withering stage, moderate at the full ripe stage and lowest at the turning stage. At the turning stage, fruits were generally most pungent except in some cultivars which were most pungent at the full ripe stage or at the withering stage.

For vegetable purpose, fruits are harvested at green immature stage. Under well managed situation, 10-12 pickings can be taken as green chilli.The quality of green chilli is decided by its size, colour and plumpness of fruit. Long, green,smooth, shiny and moderately plump fruits are preferred. For pickles, picking is done both at green and ripe stage.

Red mature fruits are harvested for dry chillies used for spice purpose. Stagered harvesting at different dates after transplanting had significant influence on fruit yield and quality. Harvesting in chilli extends from three to seven months after transplanting or direct sowing as the case may be. Ripe fruits are handpicked at frequent intervals, because retaining the fruits for a long period on plants causes wrinkles and fading of colour. Harvesting of chillies is a labour intensive operation and is carried out mostly by women labour. *Capsicum annuum* required 35-46 days to reach maturity, and showed non-climacteric behaviour, unlike fruits of *C. Baccatum*. Generally 6-8 pickings are taken under irrigated conditions and 3-4 pickings under rainfed conditions. Mature fruits after colour break stage are harvested and heaped for 1-2 days for uniform ripening and colour development.

The suggested alternative for manual harvesting is mechanical harvesting. For cost effective machine harvesting the correct plant spacing and determinate cultivar with large pendulous fruit need to be selected. Double-helix-type mechanical harvester can be used. Ethephon when applied 3 weeks before harvest at 1.75 litres ha^{-1} to promote uniform ripening, has been reported to cause considerable fruit loss affecting yield. Harvest efficiencies of 73-83 per cent could be achieved in mechanical harvesting depending on the cultivars. Trash content of the harvested chilli varies from 25 to 42 per cent of the dry weight.

5.3 Handling

Green chillies should be kept in cool dry place immediately after harvesting. Tight packing of fruits should be avoided as it encourages build up of temperatures inside the containers thus causing physiological damage to fruits. The fruit temperature at harvest may rise up to 10 °C above the ambient temperatures even under shade if tightly packed and stored. A few hours of such exposure can make the fruits to soften and show signs of shrinkage, reducing the shelf life by half.

Freshly harvested fruits are graded and usually cleaned with chlorine water (150 ppm Sodium Hypochlorite) maintained at a pH 6.5. The cleaned fruits are shade dried before packing for removal of surface moisture. Fruit should be firm, well-shaped, and free from damage caused by scars, sunburn, disease, insects, or mechanical injury before packing. Fruits that are not in shape, cracked, damaged, decayed, or wilted are discarded while grading. All fruit packed in the same carton should be similar in appearance. Then they are packed in units as desired for bulk packing. They may also be packed in to consumer packs. The produce is then pre-cooled in a cooling unit for a few hours and transferred to cold-storage ready for export.

Freshly harvested green chillies should be handled carefully while transferring to field containers to avoid damage by bruising and punctures. Containers or sacks used for picking should have smooth inner surfaces to avoid skin damage or fruit punctures. Wide, shallow, and stackable containers are advantageous as they cause less bruising of the fruit at the bottom of the container. Plastic crates with good ventilation are generally used for this purpose. Field containers should be washed with water to remove previous debris and sanitized with a sanitising solution consisting of 200 ppm sodium hypochlorite (common bleach) at a water pH of 6.5 by soaking for two minutes. Troughs or baskets made up of local materials like wood or straw can also be used. In such cases, providing a newspaper lining inside minimizes damage due to abrasion while handling and transit. For local markets medium sized sacks made of gunny or nylon may be used. Large sized sacks must be avoided as they hold large volume thus restricting aeration and damaging fruits due to heavy weight.

5.4 Grading

Grading of fresh chillies is done mainly with manual labour. *E. coli* contamination the most feared one in many countries warrants minimum human contact and maintenance of hygiene in chilli processing particularly during grading. Grading of fresh chillies is relatively easier when compared to dry chillies. Proper grading of green chillies facilitates efficient filling and canning in processing industry. The performance of an automatic Jalapeno chilli grader by three different width sizes was evaluated where the conveyor uses baby suckers to align each chilli during sensing. Chilli width was determined by means of a photodiode scanner, which detects the incoming radiation sent by a laser line generator. Chillies with necrosis are detected with a radiometer and removed to increase the product quality. The accuracy on the necrosis detection and width classification was of 96.3 and 87 per cent, respectively.

Grading is done during and after drying. Colour, glossiness, size, moisture content, pest and disease infected fruits and whitened fruits are some of the important factors that traders consider when they buy the produce from farmers. But, it is advisable to do grading at various stages starting from harvesting to prevent microbial contamination and for quality of end produce. Farmers grade the produce according to the buyer's preference. The Indian chillies are graded mostly by farmers on the basis of colour and size, before they are brought to the market. Grading at farmers' level involves separation of all visible foreign matter, discoloured pods (whitened), under sized and curly pods. The damaged, discoloured and immature pods are removed depending on market demand. Fungus infected pods are separated so that contamination in subsequent processing and storage could be avoided. Depending upon the market requirement grading is performed either during drying or after drying. However, at traders' level the most important specifications for trade are moisture and stalks. Excess moisture adds weight to the pods but gives room to various fungi to grow and subsequent development of aflatoxins. Similarly, if the stalk of the pods is broken, exposing the seeds entirely, the seeds may fall out. On the other hand in absence of optimum moisture, the pods may break and let off the seeds. Thus the seed to pod ratio in a lot, is also a valuable parameter of grade.

For different purposes, various varieties of chillies are chosen by the end user. End users are mainly of two types, such as domestic retail users and industrial wholesale users. Domestic users prefer all varieties for different occasions while industrial users who prepare chilli powder give preference for colour, pungency, thickness of pericarp and seed content. There are several local and conventional grades followed by the farmers, village merchants and itinerant merchants. Traders usually depend on the visual assessment of lots/heaps by picking handful of pods to decide upon the grades and assess theprices both in open and closed auction. The grade specifications as per the American Spice Trade Association (ASTA) is mentioned in Table 5.

Table 5: ASTA Grade Specifications (Physical and chemical) for Whole Chillies/Pods

Specification	*Suggested Limits*
ASTA Cleanliness specification	
Whole dead insects, by count	4.0
Maximum excreta, by mg lb^{-1}	1.0
Other excreta, by mg lb^{-1}	8.0
Mold, per cent by weight	3.0
Insect defiled, infested, per cent by weight	2.5
Extraneous, per cent by weight	0.5
FDA DALs	
Insect infested or mouldy pods by weight	Ave. of 3 per cent
or	
Mammalian excreta per pound indentified as to source when possible	Ave. of 1 mg
Volatile oil	N/A
Moisture	11.0 per cent max.
Ash	8.0 per cent max.
Acid-insoluble ash	1.0 per cent max.

5.5 Pre-cooling

After harvest, pre-cooling can be employed to remove field heat from the fruit and slow down metabolism. It is an important step in cold chain management. For chilli produced in the tropics, rapid cooling to 10 °C at high RH would be sufficient. Pre-cooling can be done using forced air, hydro cooling, or vacuum cooling. Hydro cooling is simpler and cheaper than the other methods, but care should be taken to prevent development of decay from inadequate drying after treatment. Pre-cooling is advisable prior to cold storage. As the shelf life of fresh chillies is very short, they are usually transported by air. The packages are stored or transported as a pallet load. To maximize savings in distribution costs, pallet loads should be established as soon as possible in the distribution chain and remain unbroken as long as possible.

5.6 Packing

Both fresh and dry chillies are usually packed at farm level to ensure easy transport, to protect the produce from any damage during storage, transportation and marketing. Good packing of chilli not only facilitates convenience in transportation and storage but also attracts consumer to pay more. The packing reduces the marketing cost and protects the quality.

Freshly harvested green chillies are generally packed in gunny sacks, bamboo baskets, or polyethylene sacks and transported to nearby domestic markets (Figure 6). For long distance marketing within a country, they are packed in plastic trays, and transported. Generally, 3-4 inches length chillies are preferred for export. Green chillies from India are exported in Corrugated Fibre Board (CFB) boxes (Table 6). Fruits are packed in consumer packs of LDPE or PP (polypropylene). Twenty consumer packs of 250 gm capacity are placed in 2 x 2 x 5 pattern in a 5 kg CFB box for export purpose.

Table 6: Specification Details for CFB Boxes for Packing and Export of Green Chillies

Sl.No.	*Specification*	*Slide Type*	*Ring *Flap Tuck-in-Type*	*RSC (Regular Slotted Container)*	*Tray with LID*
5 kg Box (Dimension: 450 x 265 x 110 mm)					
1.	Material for construction	5-ply CF	5-ply CFB	5-ply CFB	5-ply CFB
2.	Grammage (g/m²) (outer to inner)	*230X140X 140X140	*230X140X 140X140	*230X140X 140X140	*230X140X 140X140
3.	Bursting strength kg/cm²	Min. 10.00	Min. 10.00	Min. 10.00	Min. 10.00
4.	Puncture resistance inches/	Min.250	Min.250	Min.250	Min.250
5.	Compression strength Kg.	Min.350	Min.350	Min.350	Min.350
6.	Cobb (30 minutes g/m²)	Max.130	Max.130	Max.130	Max. 130

* Outer ply of white duplex board.

Source: Postharvest Manuals on Exports of Fruits, APEDA, New Delhi.

In developed countries chillies are packed in polyethyene films in unit quantities, transported using a vegetable cold chain and marketed. The effect of modified atmosphere packaging (MAP) on the postharvest quality characteristics and biochemical properties of green chilli peppers (*Capsicum annuum* L. cv. Legon 18) resulted in significantly low weight loss, retained of green colour, less amd in firmness at both 10 °C and 26-34 °C (ambient) temperatures.

5.7 Storage

Both quantitative and qualitative losses atextremely variable magnitude occur at all stages of the postharvest system. Maximum losses have been observed at the wholesale level.

The green chillies or ripe red chillies are perishable in nature and can be stored for 3 to 4 days under ambient condition without much appreciable damage to appearance and quality. However, storing them beyond this period invites mold attack, physiological deterioration, quality loss and as a result these quickly loose marketability.

The fruits of King chilli (*Capsicum chinense* Jacquin) grown under foothills of Nagaland when treated with calcium chloride (0.5 per cent), packed in non-perforated bags and stored under refrigeration at 5 °C recorded the lowest physiological loss in weight (2.55 per cent) change in fruit colour (10 per cent), highest value of ascorbic acid content (120.5 mg 100 g^{-1} of fruit) and longest shelf life (9 days) at 9th day after storage. CFTRI, Mysore recommends storing of fresh chillies at 10 ± 2 °C which extends shelf life for 10-12 days. Pre-packaging in polymeric film or paper bags improves the storage life and retains the freshness of chillies. Fresh chillies stored in polyethylene (PE) bags at 20 °C recorded a weight loss of up to 25-40 per cent accompanied by significant shrivel. The weight loss is associated with reduction in firmness of pericarp which is better at lower temperatures.Fresh green chillies can be stored up to three weeks without much postharvest losses when maintained at 10 °C (R.H. 80-90 per cent).

6.0 Domestic Marketing and Export

For international trade the ISO has specified two standard dimensions of pallets as the primary handling unit size *viz.* 1200 x 1000 mm, 1200 x 800 mm. The above two pallet sizes are used for distribution within Europe as well as for transportation of produce arriving unpalletised from European seaports and airports to its European destination. The trade channel involves several members' *viz.* a village level trader, commission agent, wholesaler, retailer, agents for exporters and exporters. The chillies have a well established spot market in India at Guntur, Warangal, Khammam in Andhra Pradesh and Raichur, Bellary in Karnataka. Guntur is Asia's largest market for chillies. The marketing season here begins in the first week of February, peaks during the month of April and closes by the middle of May. In addition to whole chilli, the marketing is also done in the form of stemless chilli and crushed chilli. For marketing, chillies are graded as first or second sort, mixture, etc. Grades such as special, medium and fair are also adopted. Important markets and feeder areas of chillies in India is presented in Table 7. Some of the important quality parameters of fresh chilli are mentioned in Table 8. Important traits of chilli for different markets are given in Table 9. Codex standards prescribe pungencyexpressed in Scoville Heat Units (SHU) in to four categories *i.e.* mild (900- 1999 SHU), medium (2000-19999 SHU), hot (20000-100000) and extra hot (> 100000 SHU).

7.0 Conclusion

Chilli, an important vegetable and spice, is used as a culinary supplement essentially for its colour, flavour and pungency. In recent years, its importance as a source of natural food colour and valuablepharmacological compounds has grown beyond imagination. There are specific varieties in Europe and South America suitable for different processing purposes, which are yet to be identified in Asia, the largest producing and exporting region. In chilli harvesting is highly labour intensive. Development of varieties suitable for mechanical harvesting and stripping will be more advantageous. Postharvest losses of vegetable chilli are very high and

Table 7: Important Markets and Feeder Areas of Chillies in India

State	Market	Feeder/catchment areas
Assam	Nagaon	Kaliabor, Dhing, Samaguri, Raha, Jamunamukh, Murajhar, Lanka and Lumding
Meghalaya	West Garo Hills	Phulbani, Rongamachokgiri, Rabbigiri, Kherapara, Rongohugui, Selsella, Tura, Batasing and Dalu.
Andhra Pradesh	Guntur	Tangeda, Dachepalle, Piduguralla, Sattenapalli, Narasaraopet, Vinuknoda, Medikonduru, Krosuru, Tulluru, Pedakurapadu, Ponnekallu and Prattipadu
Gujarat	Jamnagar	Jodia, Kalavad, Jamjodhpur, Bhavad, Kalyanpur, Dwarka, Khambalia and Dhrol.
	Banskantha	Tharad, Dhabhar, Devgani, Vodgam, Danta and Vav
Manipur	Thoubal	Yairipok, Wangjing, Wabagai and Kakchig.
	Charchandpur	Hanglep, Songsan, Tinaong, Senvon, Hanship, Molnom, Thinghat, Mulanil and Hangtam.
Maharashtra	Nasik	Kalvan, Peint, Igatpuri, Sinnar, Niphad, Yeola, Nandgaon, Satara, Furgana, Dindori and Melgaon
	Pune	Junnar, Ambegaon, Ghod, Rajgurunagar, Wadgaonsirur, Mulshi, Welhe, Purandhar, Bhor, Baramati, Indapur, Daund and Saswad.
Karnataka	Bijapur	Indi, Sindgi, Basavna Bagevadi, Muddebihal and Tikota.
	Belgaum	Athni, Arkali, Chikodi, Mukeri, Bailhongal, Ramdurg and Khauapur.
	Bangalore	Amekal, Sonnenahalti, Kannur, Bagalur, Nagarur, Marangondahalli, Haralur, Mantapa, Solurur and Chandapur.
Tamil Nadu	Thiruchirapalli	Turaiyur, Thottiyam, Musiri, Manachanallur and Lalgudi
	Coimbatore	Muttuppalaiyam, Avinashi, Tiruppur, Palladam, Udumallaippettai, Pollachi and Valparai.
	Salem	Mettur, Yercaud, Idappadi, Omalpur, Attur, Gangavalli and Sankagiri.

Table 8: Scoville Heat Units of Major International Chilli Varieties

Chilli Type	Species	Scoville Heat Units (SHU)
Ancho and Pasilla	*C. annuum* L.	1,000–2,000
Bell/sweet	*C. annuum* L.	0–100
Cascabel and Cherry	*C. annuum* L.	1,000–2,500
Cayenne	*C. frutescens* Roxb.	30,000–50,000
De Arbol	*C. annuum* L.	15,000–30,000
Espanola	*C. annuum* L.	1,000–1,500
Guntur Sannam	*C. annuum* L.	30,000–40,000
Habanero	*C. chinense* Jacq.	200,000–350,000
Jalapeno and Mirasol	*C. annuum* L.	2,500–5,000
New Mexican	*C. annuum* L.	500–1,000

Contd...

Table 8–*Contd...*

Chilli Type	*Species*	*Scoville Heat Units (SHU)*
Serrano	*C. annuum* L.	5,000–15,000
Scotch Bonnet and Thai	*C. chinense* Jacq.	100,000–350,000
Tabasco	*C. frutescens* Roxb.	50,000–100,000
Chiltepin	*C. annuum* var. *glabriusculum*	
Tezpur/Nagarhari (Assam)	*C. chinense* Jacq.	8,55,000-1 million

Table 9: Important Traits for different Markets

Market Type	*Important Traits*
Fresh whole fruit market	Colour (light green, dark green, yellow, orange and multicolour), size (2-15 cm), shape (cherry or bell type or short stout to long slender), pungency, flavour, exocarp thickness (both thin and thick), endocarp to seed ratio, seed colour, vitamin C and A.
Processing of fresh chillies	Colour, pungency, flavour, size, shape, pericarp thickness, endocarp to seed ratio, suitability for processing and shelf life after processing.
Dried chillies (whole and powdered)	Colour, pungency, flavour, dry weight, crude fibre, endocarp to seed ratio and seed content.
Oleoresin	Colour, pungency (nil to extra hot) and essential oil.
Capsaicin	Extra hot (> 100000 SHU) pungency and recovery.
Capsanthin	High EOA colour (above 50000 EOA) value and recovery.
Ornamental	Colour (multi colour and variegated), fruit shape, size, erect and clustered fruits.

development of cold chain technology keeping in mind the sensitivity and chilling injury of the crop under low temperature storage is needed. Under ambient conditions, the storage life of fruits can be extended by packing them inperforated polythene film. Global trade of chilli and its value added products is affected by trade rejections due to the presence of adulterants, pesticide residues and aflatoxins. Adoption of Good Agricultural Practices (GAP) must be promoted among the producers to deliver a quality and hygienic produce to the consumers.

8.0 Important References

Bosland, P. W., Votava, E. J. and Votava, E. M. 2012. Peppers: vegetable and spice capsicums (No. 22). CABI.

Deepa, G. T., Chetti, M. B., Khetagoudar, M. C. and Adavirao, G. M. 2013.Influence of vacuum packaging on seed quality and mineral contents in chilli (*Capsicum annuum* L.). *Journal of Food Science and Technology (Mysore)*. **50(1)**: 153-158.

Edusei, V. O. Ofosu-Anim, J. Johnson, P. N. T. Cornelius, E. W. 2012. Extending postharvest life of green chilli pepper fruits with modified atmosphere packaging. *Ghana Journal of Horticulture*. **10**: 131-140.

FAOSTAT. 2014. At: http: //faostat.fao.org/[Accessed on 02 June 2014].

Hahn, F. 2002. Automatic Jalapeno chilli grading by width. *Biosystems Engineering*, **83(4)**: 433-440.

Jansasithorn, R., East, A. R., Hewett, E. W., Molan, A. L., Heyes, J. A., and Mawson, A. J. 2010. Harvest Maturity Influences the Antioxidant Activity in Jalapeño Chili. In: *XXVIII International Horticultural Congress on Science and orticulture for People (IHC2010): International Symposium, August, 2010*. pp. 379-384.

Mendez, M.A., Ligarreto, G.A., Hernandez, M.S. and Melgarejo, L.M. 2004. Evaluating growth and determining harvesting index in four types of hot chilli pepper (*Capsicumsp.*) grown in the Colombian Amazonian Region. [Spanish] *Agronomia Colombiana*. **22(1)**: 7-17.

Pankar, D. S. and Magar, N. G. 1978. Capsaicin, total colouring matter and ascorbic acid contents in some selected varieties of chilli (*Capsicum annuum* L.). *Journal of Maharashtra Agricultural University*, **3(1)**: 116-119.

Pruthi, J.S. 1993. Major spice of India - Crop management and postharvest technology. ICAR. pp. 221-222.

Raina, B.L. and Teotia, M.S. 1986. Evaluation of chillies (*Capsicum annuum* L.) grown in Jammu and Kashmir. *Indian Food Packer*. **40(5)** 6-10.

Rongsennungla, I., Kanaujia, S. P., Singh, V. B., and Maiti, C. S. 2012. Effect of pre- and postharvest treatments on yield, quality and postharvest shelf life of king chilli (*Capsicum chinense* Jacquin) under foothill conditions of Nagaland (India). *Journal of Spices and Aromatic Crops*. **21(1)**: 42-47.

Shivhare, U.S., Narain, M., Agarwal, U.S., Saxena, R.P. and Singh, B.P.N. 1987. Some physico-chemical properties of chillies. *Journal of Food Science and Technology*. **24 (2)**: 98 99.

Taufik, M. 2011. Farming analysis and postharvest handling of red chili. [Indonesian] *Jurnal Penelitian dan Pengembangan Pertanian*. **30(2)**: 66-72.

Teotia, M.S. and Raina, B.L. 1987. Evaluation of chillies grown in Punjab. *Indian Food Packer*. **41(3)**: 81-86.

Uma Jyothi, K. and Ravi Shankar, C. 2006. Effect of postharvest chemical treatments on storage of dried chilli cv. LCA 334. *Vegetable Science*. **33(1)**: 38-44.

Wall, M. M. Walker, S. Wall, A. D. Hughs, E. Phillips, R. 2003. Yield and quality of machine harvested red chilli peppers. *HortTechnology*. **13(2)**: 296-302.

Chapter 29
Cut Flowers

Kushal Singh

Senior Plant Physiologist

Department of Floriculture and Landscaping, Punjab Agricultural University, Ludhiana – 141 004, Punjab, India

Ramesh Kumar

Director, Directorate of Floricultural Research, ICAR-Indian Agricultural Research Institute, New Delhi – 110 012, India

E-mail: kushal.flori@pau.edu; E-mail: directordfr@gmail.com

1.0 Introduction

Cut flowers are an important component of domestic as well as international floriculture trade. Cut flowers are used for decoration in vases, in flower arrangements, in bouquets, for decoration at auspicious occasions and as wreaths. The demand for cut flowers in the global market is increasing at the rate of 10-15 per cent each year. Important flowers which dominate the global trade include rose, carnation, chrysanthemum, orchids, tulip, gerbera, anthurium, lilies, gypsophila, bird of paradise, alstroemeria etc.

The cut flower trade is highly competitive and requires high quality product. Since flowers are highly specific in their temperature requirements, most cut flower crops are cultivated under protected conditions where the temperature conditions and nutrition is precisely controlled. Due to increase in labour costs and freezing temperatures during winter, the cultivation of cut flowers in Western Europe has become cost intensive. As a consequence, the cut flower production has shifted to the developing countries.

The varied climate of India is highly suitable for growing cut flowers throughout the year through the selection of the flower type and agro-climatic zone in the country.

2.0 Status of Cultivation

As per the NHB database on horticulture, the total area under cut flowers in India was 2.32 lakh hectare with total production of 767.31 lakh MT in the year 2012-13. The leading cut flower producing states are West Bengal (33.1 per cent), Karnataka (12.3 per cent), Maharashtra (10.3 per cent), Andhra Pradesh (9.0 per cent), Odisha (7.9 per cent), Uttar Pradesh (6.4 per cent), Uttarakhand (4.7 per cent), Assam (4.9 per cent) Himachal Pradesh (2.3 per cent), Jharkhand (2.2 per cent) and others (6.8 per cent).Maximum production of cut flowers in India comprises of rose followed by tuberose, gladiolus, gerbera and carnation.

3.0 Major Commercial Varieties

The important floricultural crops in the international cut flower trade are rose, carnation, chrysanthemum, gerbera, gladiolus, gypsophila, liastris, nerine, orchids, achillea, anthurium, tulip, and lilies. Floriculture crops like gerbera, carnation, rose, lilly etc. are grown in green houses etc are open field crops while chrysanthemum, roses, gaillardia, lily marigold, aster, tuberose etc.Postharvest lasting quality of flower species and cultivars vary considerably. For longer postharvest life, selection of variety is considered very important. Differences in flower longevity and quality among the varieties may be due to differences in their structural and physiological make up. For instance, *Alstroemeria* cv. Rosaria exhibits vase life of 17 days whereas cv. Pink Panther shows vase life of 8 days. Likewise, cv. Marleen of gerbera shows vase life of 20.5 days in comparison to cv. Agnes with vase life of 8.3 days. Similar variations in vase life have also been reported in spray carnation, rose, *Anthurium*, *Lilium*, rose and gladiolus and fronds of *Asparagus* and *Nephrolepis*. The flowers belonging to families Orchidaceae and Asteraceae possess long vase life whereas flowers like iris, tulip and narcissus have short vase life. Not only this, the varieties grown under protected conditions generally possess higher vase life than those grown under open field conditions because the climatic factors and nutritional requirements of the crop are better managed in the green house. Greater variations are also reported to occur in vase life due to variations in climatic conditions, texture and pH of the soil.

4.0 Nature and Causes of Postharvest Losses

Factors involved in the decline of vase life of flowers include loss of carbohydrates due to high respiration rates, water deficit conditions which are induced when the flowers are handled dry during postharvest handling, sensitivity to ethylene, incidence of pests and diseases, quality of irrigation water and pollutants present in the atmosphere or irrigation water. When detached from the plant the cut stem is deprived of its natural sources of food, water, minerals and hormones and carriers out its metabolic functions as an independent entity at the expense of its own available resources. Postharvest management involves steps taken for the improvement of flower vase life so that the consumers can get the flower in fresh condition. Before reaching the consumer, the cut stem passes through many handling steps such as cooling, grading, treatment with preservatives, storage, packing and transit.

4.1 Water Relations of Cut Flowers

The wilting of petals of many cut flowers leads to the termination of vase life even though the stems are constantly placed in water. After cut stems are placed in water after harvesting, they undergo marked changes in fresh weight which initially increases due to bud development and then shows decrease due to senescence of the foliage and the flower. The stems with large leaf area like those of rose and chrysanthemum are highly prone to water loss. In general, cut stem continues to exhibit decline in water absorption with time. As water loss exceeds water absorption, the stem develops water deficit conditions leading to wilting of petals as well as leaves. The reduced uptake of water by the stem has been reported to be mainly due to plugging of xylem vessels by the growth of bacteria, fungi and other microorganisms in vase water or on the stem surface. Microbial growth causes an increase in the stem resistance to water transport via xylem channels and decrease in vase life. The metabolites produced by bacterial growth have been found to reduce longevity. The cut surface of a flower stem also releases proteins, amino acids, sugars and minerals into the vase water which provides an ideal food for the growth of bacteria, yeasts, and fungi. Decrease in water absorption by the stem and stomatal loss of water also leads to development of air-filled spaces (emboli) within the xylem, which cause cavitation and hinder the uptake of water by the stem. Emboli can, however, be reduced by re-cutting 2-3 cm bases of the stems under water, acidification of the solution to pH 3 to 4 or by placing the stem neck deep in water. Xylem occlusion also occurs by ethylene produced at the cut end of the stem, wounding reaction that involves both peroxidase and catechol oxidase activity and also deposition of lignin, suberin and tannins. Water deficit conditions are known to cause physiological disorders in the cut stems such as stem break in gerbera and bentneck of rose. Adverse water relations also cause changes in hormonal balance.

Quality of water also considerably affects keeping quality of the flowers and the foliage. Hard water usually contains minerals that make water alkaline. High pH of water reduces water uptake by the stem. Total dissolved solutes (TDS) present in water also affect vase life of the flowers. Total dissolved solutes more than 700 ppm decrease vase life of gladiolus whereas at low concentration *i.e.* at 200 ppm they adversely affect vase life of carnation, chrysanthemum and rose. Tap water also contains fluorides which are toxic to many flowers such as rose, gladiolus, carnation and freesia. Cut roses are also sensitive to sodium ions. Deionized water, boiled tap water, distilled water or millipore-filtered water are considered appropriate for cut flowers. Acidification of water (pH 3 to 4) also helps to prolong longevity of flowers as low pH decreases the bacterial population and increases the rate of water flow in the stem.

5.0 Postharvest Management

5.1 Preharvest Factors

Postharvest behavior of cut flowers is determined by the preharvest conditions under which the crop is grown. The preharvest conditions which have an important bearing on vase life of flowers are: selection of the variety, environmental factors *viz.*

light, temperature and relative humidity, nutrition, irrigation, diseases and pests, pollutants etc.

It is estimated that postharvest keeping quality of the flowers is determined by the preharvest growing conditions which account for 30-70 per cent of vase life.

5.1.1 Environmental Factors

Duration as well as intensity of light and temperature play a major role in determining the quality of the flowers. Most cut flower crops requires well-lighted conditions *i.e.* about 3200 ft candle of light intensity for optimal growth and flower production. Low light conditions cause decrease in photosynthetic efficacy of leaves and result in flowers with short vase life.Reduced light intensity can also lead to decrease in anthocyanin content of petals and increase in bluing of petals in red cultivars of rose. Besides, low light intensities lead to weak stems and increased susceptibility to bent neck in rose. Very high light intensities accompanied by high temperatures and increased ethylene liberation cause abscission of flowers/petals. High temperatures also increase loss of water from the plants and also increase the respiratory activity which depletes the flowers of their carbohydrate reserves. Flower crops require moderate and specific temperatures for growing. For instance, rose requires 20-25 °C day and 16 °C night temperature whereas chrysanthemum requires 20-28 °C day and 15-20 °C night temperature for flower bud initiation. Carnation, being a cool climate crop requires 20 °C day and 10 °C night temperature for optimum growth and flower production. Low temperatures below 10 °C cause retardation in plant growth and sub-zero temperatures can cause freezing injury to the buds. As a consequence, most flower crops are grown under protected conditions where optimum light and temperature conditions are precisely controlled. High relative humidity in the atmosphere is highly suitable for germination of fungal spores and hence, can cause incidence of diseases. Incidence of *Botrytis* blight increases manifold when high relative humidity is accompanied by low temperatures. Therefore, relative humidity should be moderate and near 60-80 per cent for raising a good crop.

5.1.2 Nutrition

Flower crops require optimum supply of nutrients for sustained and quality production of flowers.Therefore, cut flower crops like rose, lilies, carnation, gerbera etc. are grown under green houses where the nutrients are provided through drip irrigation system. Excess of nutrients cause increase in pH and adversely affect their uptake. Besides, high salt concentration in the root zone causes inhibition of root growth. Especially, application of high nitrogen doses should be avoided as they result in undesired longer and weaker stems and high incidence of diseases. The nitrogen deficient crop, on the other hand, produces pale green foliage and blooms of poor quality. Deficiency of micronutrients, especially those of iron, manganese and zinc is of common occurrence in many cut flower crops and should be corrected before they can cause any serious damage.

5.1.3 Irrigation

Irrigation schedule of flower crops varies with the species and prevailing climatic conditions. Excessive irrigations cause water-logged and anaerobic conditions which inhibit both root as well as shoot growth leading to production of poor quality flowers. Water-logged conditions also invite pathogenic fungi, especially gray mold (*Botrytis cinerea*).

5.1.4 Diseases and Pests

Flower crops are attacked by a variety of pathogens and pests. Especially, under polyhouse conditions, the incidence of spider mite is a major problem. The flowers damaged by pathogens, insects and pests show poor vase life and result in rejection in the market place. Damaged tissues are potential sources of ethylene which causes senescence of flowers and foliage on the stem. It is, therefore, essential to keep the crop free from diseases and pests. Sanitation of growing areas, especially greenhouses, grading rooms and cold store is very essential to keep flowers free from diseases during production as well as after harvest. Especially, green houses should be kept clean and ventilated to allow free circulation of fresh air.

5.1.5 Pollutants

The flower crops should not be grown close to the industries which release toxic gases like sulphur dioxide, nitric oxide and fluorides in the atmosphere. Sulphur dioxide and nitric oxide form weak acids when come in contact with moisture and cause discolouration of petals. Tip burn of leaves commonly observed in gladiolus is caused by high concentration of fluorides in the atmosphere and/or irrigation water. The fluoride accumulate in the tip regions of the leaves and cause burning of leaf tips.

5.2 Harvesting

Proper harvest maturity, time and techniques of harvesting the flowers determine postharvest vase life of flowers to a greater extent. Flowers should always be harvested at an optimum maturity stage. Too immature buds fail to open whereas mature flowers have short vase life. The stage of harvest varies with the species and even with the cultivar, prevailing temperature conditions, marketing system and requirement of the consumer. In general, for transport to the distant markets, the flowers should be harvested at bud stage whereas for the local market, the flowers are harvested at relatively open stage.

5.2.1 Stage and Time

Flowers like rose, carnation, gladiolus, daffodils, lilies, iris, freesia and tulip can be harvested at bud stage as buds of these flowers possess the ability to open after storage, transport and distribution. On the other hand, flowers like orchids, dahlia, gerbera, calendula etc. are harvested when fully open. The optimum stages of harvest for some important flowers are given in Table 1.

Bud harvesting of flowers is always beneficial as the buds continue to open over a longer time and hence, possess long vase life. Besides, the buds are less sensitive to ethylene, easy to handle during storage and transport, occupy less space in packing boxes and are hence, economical to transport. Besides, the buds are less prone to damage by diseases and pests.

Table 1: Stages of Harvest for Important Cut Flowers

Flower Name/Species	*Stage of Harvest*
Alstroemeria (lily of the Incas)	4-5 florets open
Anthurium andreanum	spadix almost fully developed
Antirrhinum majus (snapdragon)	1/3 florets open
Bouvardia hybrids	Flowers start opening
Calendula officinalis	Fully open flowers
Callistephus chinensis	Fully open flowers
Cattleya spp. (orchid)	3-4 days after opening
Chrysanthemum morifolium	
(Florist's chrysanthemum)	
Standard	When outer florets fully expanded
Spray	Flowers open but before shedding of pollens
Pompons and decorative	Centre of the oldest flower fully open
Anemones	open but before central disc florets begin to elongate
Cyclamen persicum	Fully open flowers
Cymbidium species (orchid)	3-4 days after opening of all flowers
Dahlia variabilis	Fully open flowers
Dendrobium spp. (orchid)	Fully open flowers
Dianthus caryophyllus (carnation)	
Standard	Paint brush stage when flowers are half open
Spray	At least 2 flowers fully open
Freesia hybrids	Coloured buds
Gerbera jamesonii	Flowers open but outer 2 rows show shedding of pollens
Gladiolus cultivars	1-5 florets show colour
Gloriosa superba	Fully open flowers
Gypsophila paniculata	Flowers just open
Hemerocallis spp.	Coloured buds
Hippeastrum hybrids	Coloured buds
Iris germanica	Coloured buds
Lilium spp. (lily)	Coloured buds
Narcissus spp. (daffodil)	Goose neck stage
Nerine bowdenii	First bud almost open
Ornothogalum umbellatum	Coloured buds
Papaversominiferum	Coloured buds
Paphiopedilum spp.	3-4 days after opening
Polianthes tuberosa (tuberose)	
Single	Buds fully developed but unopened
Double	Maximum florets open
Rosa hyrida (Rose)	1-2 petals beginning to unfold
Sterlitzia reginae (bird of paradise)	First floret opens
Tulipa spp. (Tulip)	Half coloured buds
Zinnia elegans	Fully open flowers

Flowers should always be harvested when temperatures are mild because high temperatures increase respiration and also promote water loss. Flowers should, therefore, be harvested either in the morning or in the evening. In the morning, flowers are fully turgid due to low transpiration rate that occurs during night. Flowers such as rose, chrysanthemum and gerbera which are highly prone to water stress benefit when harvested in the morning. Evening harvest is also advocated because of high sugar level in the stem due to increased activity of photosynthesis during day time but due to high evening temperatures, these flowers are more prone to desiccation. After harvest, the stems should be immediately placed in water or preservative solutions to prevent water loss.

5.2.2 Method

Proper method should be adopted to harvest the flowers. The stems should always be cut with sharp tools (knives or secateurs). Specialized tools are also used to harvest the flowers. For example, rose shear which grips the stem after it has been harvested is used to harvest roses. While harvesting, care should be taken not to crush the stem. Hardwood stems should always be given slanting cut so as to expose the maximum surface area to ensure rapid water absorption.

Specific guidelines should be followed for harvesting of flowers. For instance, stems of chrysanthemum should be harvested at least 10 cm above the ground level. The stem close to the soil is hard and lignified. The lignified tissues hinder the uptake of water by the cut stem. The stems of rose should be harvested above at least two leaves (with five leaflets) leaves to ensure sufficient photosynthetic area on the plant for sustained production of flowers. Cutting the stem near the base may also result in blind shoots in the subsequent season.

Stems of some species like dahlia, poinsettia and papaver release latex upon cutting. Such stems should be given dip in hot water (80-90 °C) for few seconds to inactivate the latex-releasing tissues. Immediately after harvest, the flowers should be placed in water or preservative solution and shifted to cold rooms to remove field heat generated due to high rate of respiration. Under no circumstances, the flowers should be placed out of water for more than few minutes.

5.3 Cooling and Conditioning

Immediately after harvesting or when the flowers are received from far off places, they are required to be brought to low temperature conditions. This will help to remove the field heat generated due to high rate of respiration, check the excessive opening of the flower bud, reduce water loss and production of ethylene from the flowers.

In many instances, the flowers and foliage after prolonged dry storage as well as transportation show signs of wilting due to excess loss of water and hence, require restoration of turgidity. Such stems are needed to be rehydrated by placing them in solution of biocide such as aluminium sulphate or citric acid (3 g/litre of water) or in water acidified to pH 3-4. This process of rehydrating the stem is referred to as "conditioning". In case of severe wilting, the stems should be placed neck deep in water. Before placing the stems in hydrating solution, should be given re-cut at the base to expose fresh tissues for improving water uptake. These should then be placed at cool temperatures (2-4 °C) to prevent further loss of water.

5.4 Floral Preservatives

Term "Floral Preservative" is used for any chemical formulation which is used to extend the vase life of flowers. The use of chemical formulations to extend vase life has been an integral part of postharvest management of cut flowers since long.Many commercial floral formulations are now available in several developed countries for improving vase life of the flowers. The chemical treatments can be used in the form of pulsing, bud opening solutions or holding solutions and the composition of each varies in different species depending upon the mode of senescence.

Floral preservatives mainly contain sucrose (as a carbon source) and a biocidal compound (that curtails microbial growth in vase water as well as on the cut surface of the stem) as basic constituents. Besides, plant growth regulators or ethylene antagonists may also form a constituent part of the floral formulations. Floral preservatives can be applied as

i) Pulsing solutions;
ii) Bud opening solutions; and
iii) Holding or vase solutions.

5.4.1 Pulsing

The term "pulsing" refers to short duration (few minutes to few hours) pretransit or prestorage treatment of harvested flowers in pulsing solution. Sugar (sucrose) is a main component of the pulsing solution. Since pulsing involves short duration treatment, high concentrations of sucrose are used.Pulsing treatment is given at moderate temperatures (20-25 °C) and well-lighted conditions to allow maximum solution uptake by the stem. Concentration of sucrose for pulsing varies with species and endogenous levels of carbohydrates in the bud. The suitable concentration of sucrose for pulsing treatment has been suggested to be 20 per cent for gladiolus and gerbera, 10 per cent for carnation, bird of paradise and gypsophila and 2 to 5 per cent for rose and chrysanthemum. Sucrose should be used in combination with a biocidal compound appropriate to the crop to prevent microbial growth. Buds in flowers like rose exhibit accelerated opening under ambient conditions. Such flowers should be given pulsing treatments at cool temperatures (2-4 °C). Since treatment with sucrose invites growth of pathogens, it is desired that the flowers which are to be exported or transported to distant markets should not be pretreated with sucrose. Instead, such flowers should be treated with biocidal solution so as to prevent microbial growth on the surface of the stem during transit.

Silver thiosulphate (STS), an inhibitor of ethylene action has found wide spread use in improving vase life of ethylene-sensitive cut flowers. However, since silver in STS is seen as an environmental pollutant, its use is not desired in the developed countries. Ethylene-sensitive flowers *viz.* carnation can, however, be treated with 1-methylcyclopropene (1- MCP) before storage or transportation.

Stems of alstroemeria and lilium can be pulse-treated with GA_3 or those of chrysanthemum with thidiazuron (TDZ) to prevent premature yellowing of leaves. Short duration pulsing with NaOCl (1 per cent) plus tween 80 (0.01 per cent as a wetting agent) for one min or with $AgNO_3$ solution (1000 ppm) for 10 min has been recommended for gerbera stems. Suitable pulsing solutions for some important cut flowers are given in Table 2.

Table 2: Suitable Pulsing Treatments for some Important Cut Flowers

Name of the Flower	*Treatment*
Alstroemeria	1-MCP, $20nl^{-l}$ for 6 h
Anthuriummajus	BA, 150 mg/l
Antirrhinum majus L.	1-MCP, $20nl^{-l}$ for 6 h
Carnation	☆ Sucrose, 10-20 per cent + Physan 20^R, 200 ppm for 16-20 h
	☆ STS, 4mM for 10 min
	☆ 1-MCP, $20nl^{-l}$ for 6 h
	☆ Sucrose, 10 per cent +STS,1mM for 20 h
	☆ 1-MCP, 100 ppb for 30 min
Chrysanthemum	☆ Sucrose, 5 per cent + Physan 20^R, 200 ppm for 16-20 h
	☆ Methanol, 4 per cent or BAP, 210 ppm for for 24 h
Gerbera jamesonii	NaOCl, 1 per cent +Tweeen 20, 0.1 per cent or AgNO3, 1200 mg/l for 10 min
Gladiolus	☆ Sucrose, 20 per cent + Aluminium sulphate, 400 for 20 h
	☆ Sucrose, 20 per cent + Physan 20^R for 16-20 h
Gypsophila	Sucrose, 10 per cent + Physan 20^R, 200 ppm for 16-20 h.
Rose	☆ Sucrose, 3 per cent + Citric acid, 320 ppm at 4 °C for 16-20 h
	☆ Sucrse, 3 per cent + Aluminium sulphate, 300 for 16-20 h
Stock (*Mathiola incana* L.)	☆ TDZ, 5mM for 24 h
	☆ 1-MCP, $20nl^{-l}$ for 6 h
Sweet pea	4mM STS for 8 min

5.4.2 Bud Opening Solutions

The immature buds of many flowers which otherwise do not open in water can be made to open in chemical solutions before they are sold in the market and are referred to as "bud opening solutions". The components of bud opening solutions are essentially the same as those of pulsing solutions, but in case of bud opening solutions, lower concentrations (2 to 5 per cent) of sucrose are used. Buds are usually opened at moderate temperatures (20-25 °C). The bud opening solutions have been standardized in many flowers like standard chrysanthemum, carnation and gladiolus The immature buds of spray chrysanthemum could be perfectly opened in a solution containing 2 per cent sucrose plus silver nitrate (25 mg/l) plus citric acid (75 mg/l) plus 6-benzyl adenine (50 mg/l). Similarly, buds of standard chrysanthemum can also be opened in a solution containing sucrose and 8-HQC though the concentration of sucrose required for the opening of buds varies with the cultivar and ranges from 2 to 30 per cent. The immature florets of gladiolus spikes cut at tight bud stage can be made to open with a solution containing sucrose (2 to 4 per cent) and aluminium sulphate (400 ppm)

5.4.3 Vase Solutions

Holding or vase solutions are used to hold flowers continuously, till the termination of their vase life. The level of sucrose in vase solutions is also kept low (0.5 to 2 per cent) due to long durations for which flowers are placed in the solution. Suitable vase solutions for some important cut flowers are given in Table 3.

Table 3: Suitable Vase Solutions for some Important Cut Flowers

Name of the Flower	*Treatment*
Carnation	STS, 4mM for 15 min f.b. Sucrose, 5 per cent + Aluminium sulphate, 200
Chrysanthemum	Sucrose, 1.5 per cent +8-HQC, 200 mg/l or Sucrose, 1.5 per cent + $AgNO_3$, 25 mg/l+citric acid, 75 mg/l
Dutch iris (*Iris hollandica*)	Sucrose, 4 per cent +Aluminium sulphate, 300 ppm
Eustoma (*Eustoma grandiflorum* (Raf.) Shinn	Sucrose, 1.5 per cent +Aluminium sulphate, 200 ppm
Gerbera (Gerbera jamesonii)	$AgNO_3$, 20 mg/l
Gladiolus	✫ Sucrose, 2 per cent + 8-HQC, 200 ppm
	✫ Sucrose, 4 per cent +Aluminium sulphate, 400 ppm
	✫ Sucrose, 4 per cent +Aluminium sulphate, 300 ppm
	✫ Sucrose, 2-4 per cent + DICA, 50 mg/l
	✫ Sucrose, 2-4 per cent + $AgNO_3$, 25 mg/l+ciric acid, 75 mg/l
Oncidium	Sucrose, 4 per cent +8-HQS, 100 mg/l
Rose	✫ Sucrose, 1.5 per cent +Aluminium sulphate, 300 ppm
	✫ Sucrose, 1.5 per cent +Aluminium sulphate, 300 ppm or chlorine solution, 50 ppm prepared from $CaOCl_2$

Water is the most important component of floral preservatives. High pH (alkalinity) and presence of total dissolved solutes (TDS) in tap water adversely affects flower vase life. High pH promotes the microbial growth and hence, reduces vase life. Therefore, acidification of water to low pH 3.0-3.5 is advantageous because it decreases microbial growth and improves water uptake by the stem. The threshold levels of TDS in water are toxic to some flowers (disussed under "water relations of cut flowers").

Sucrose is the most widely used sugar in floral preservatives though other sugars like glucose and fructose are also reported to improve vase life of flowers. This beneficial effect of sugars on flower senescence can be attributed to the supply of substrates for respiration, structural materials and osmoticum. Exogenous application of sugars supplies the flower with much needed substrate for respiration and not only prolongs the vase life but also enables cut flowers harvested at the bud stage to open, which otherwise would not occur naturally. Sugars also improve water balance of cut flowers. Besides, sugar decreases sensitivity to ethylene. The optimum concentration of sugar required, however, varies from species to species and type of treatment. Since sugar promotes microbial growth, it is usually combined with biocidal compounds before use. Important biocides used for treating cut flowers are 8-hydroxyquinoline citrate, silver nitrate, aluminum sulphate, citric acid, slow release chlorine compounds and quaternary ammonium compounds. HQC (8-hydroxyquinoline citrate) is very effective broad spectrum biocide. It acidifies water and also induces partial closure of stomata. This compound is used at concentration of 200-600 ppm. Silver nitrate ($AgNO_3$) completely inhibits the microbial growth at the concentration of 25 ppm.

Since silver, being a heavy metal is considered to be an environmental hazard, its use in cut flower industry is no longer desired. Aluminium sulphate (100-300 ppm) has also been found very effective in increasing keeping quality of many flowers and is a standard biocide used in commercial preservatives. Besides reducing pH of the solution, it inhibits bacterial growth and also causes partial closure of stomata. Since, aluminium sulphate is also reported to cause foliage wilting in chrysanthemum, it should be used with caution.

Citric acid (50-1000 ppm) has been reported to improve vase life of roses, chrysanthemum, gladiolus and carnation and is also a very effective compound for rehydration of cut stems.

Several slow release chlorine compounds such as sodium dichloroisocyanurate (DICA) and 1,3-dichloro-5,5-dimethyl hydantoin (DDMH) have also been found useful for many flowers. Chlorine solution prepared from sodium hypochlorite or by dissolving bleaching powder (calcium hypochlorite) in water has also been reported to extend vase life of flowers like gladiolus, rose and carnation.The concentration of chlorine in solution should be kept at 50 ppm because high concentrations can cause damage to the foliage as well as flower buds. For roses, only bleaching powder is recommended as sodium hypochlorite causes necrosis of the leaves and abscission of buds. Quaternary ammonium compounds have also been found very useful biocides for cut flowers. Besides, some other compounds such as thiabendazol, dichlorophen, salts of nickel, cobalt and copper and benzoic acid have also been found to extend vase life of some flowers. The biocides should be used with caution because the suitability of biocidal compounds may vary with the species.

5.5 Role of Growth Regulators

Plant growth regulators (PGRs) play a central role in regulating the senescence process of petals. Petal senescence is probably controlled by an interaction of plant growth substances rather than by a single hormone. The role played by PGRs in senescence process is of considerable use in evolving techniques to improve the vase life of cut flowers.

5.5.1 Gibberellins

Gibberellic acid is known to extend vase life of excised petals of carnation and easter lily and accelerate opening of immature buds of gladiolus. Gibberellic acid reduces ethylene production and delay senescence of cut carnation flowers. GA_3 has also been reported to prevent chlorophyll degradation in leaves of *Zantedeschia*, lilum and alstroemeria.

5.5.2 Cytokinins

Cytokinins, often regarded as senescence retarding hormones, especially in leaves are also reported to delay petal senescence. The aging of flower petals is often accompanied by a decline in the endogenous levels of cytokinins. Cytokinins have also been shown to reduce sensitivity of flowers to ethylene and also delay the ethylene climacteric. Cytokinins delay petals senescence by maintaining cellular integrity and proteins. There is an inverse relationship between cytokinin content and senescence in some flowers like petunia, roses and carnation There are also indications that cytokinins may also act as free radical scavengers in cells.

Cytoknins delay yellowing of leaves in miniature gladiolus, alstroemeria, some cultivars of chrysanthemum, asparagus and arum lily. Prestorage foliar sprays or foliar dipping with cytokinins as well as GA_3 are reported to delay yellowing of leaves though efficacy of these bioregulators varies with the species. Thidiazuron, a substituted phenyl urea compound with cytokinin like activity also retards postharvest senescence of leaves in cut flowering stems.

5.5.3 Polyamines

Polyamines which are endogenous growth regulators are known to possess anti-senescence properties. Polyamines have been reported to improve vase life of carnation and gerbera.

5.5.4 Ethylene

Senescence of many flowers is often associated with increased production of ethylene. In ethylene-sensitive flowers, ethylene is regarded as a signal, mediating a sequence of events that eventually lead to the death of petals. The typical effects of ethylene on flowers include inrolling of carnation and kalanchoe petals which indicates the first sign of senescence, fading and inrolling of corolla in *Ipomoea*, fading of sepal tips in orchids and abcission of flowers and petals. Ethylene causes premature senescence and wilting of petals in a wide range of flowers especially, those belonging to families Campanulaceae, Malvaceae, Caryophyllaceae, Scrophulareaceae and Orchidaceae. The sensitivity of flowers to ethylene depends upon the stage of bud development, concentration of ethylene and exposure time.The exposure of mature flowers to ethylene leads to autocatalytic production of ethylene associated with an increased respiration rate and consequent increase in fresh weight losses. In ethylene-sensitive flowers, pollination induces increased ethylene production and wilting or abcission of petals. Exogenous ethylene is also reported to accelerate chlorophyll degradation in some cultivars of chrysanthemum. Silver thiosulphate (STS), an inhibitor of ethylene action has frequently been used to extend vase life of ethylene-sensitive flowers. However, due to high toxicity of the silver ions included in STS which involves enviormental concern, there is a great interest in 1-MCP, which prevents the increased production of this phytohormone through permanent bonding to the ethylene receptor. Moreover, 1-MCP prevents bud/flower desiccation and abscission as well as leaf yellowing at very low concentrations.

Since pollinated flowers release considerable amount of ethylene, they should immediately be removed from the greenhouses. The greenhouses, packaging areas and storage rooms should be kept clean by removing diseased and damaged plant tissues. Flowers should be handled with care so as to avoid any physical damage. Flowers should not be stored along with fruits and vegetables as the latter produce considerable quantities of ethylene. Flowers can also be protected from ethylene by removing ethylene from the atmosphere. The greenhouses, packaging areas and the storage rooms should be ventilated to remove ethylene that continues to accumulate in these areas.

5.5.5 Inhibitors and Retardants

Abscisic acid (ABA), a natural plant growth inhibitor when applied exogenously accelerates petal senescence in carnation and increases sensitivity towards ethylene. ABA content is reported to increase with petal senescence in carnation and rose Under well-lighted conditions ABA, however, prolongs vase life of cut roses apparently by closing stomata and preventing water loss. ABA increases vase life of cut roses in low pH 6 solution due to stomatal closure but at higher pH 8 it reduces vase life, probably due to the development of microbes in vase solution at high pH. Pulse treatment with sucrose plus abscisic acid suppressed leaf damage and improved the quality of cut *Eustoma* flowers

5.5.6 Growth Retardants

Growth retardants such as daminozide and chloromequat chloride have also been reported to extend flower vase life of some flowers, though their mode of action is not fully understood.Methyl jasmonate, another naturally occurring plant growth regulator has been reported to accelerate ethylene production and senescence in flowers of dendrobium and petunia.

Nitric oxide (NO), a diffusible natural growth regulator plays an important role in alleviation of biotic and abiotic stresses and programmed cell death. Nitric oxide donor compound 2,2–(hydroxynitrosohydrazino) bisethanamine (DETA/NO) extend vase life of flowers of snapdragon, delphinium, chrysanthemum, tulip, gerbera, oriental lily, rose and iris cut stems. NO seems to have commercial potential because solid NO releasing water soluble compounds are available.

5.6 Storage of Flowers

Storage of flowers is highly advantageous to regulate the supply of flowers in the market in view of increasing problems regarding preservation of large volumes of flowers and their distribution to the consumers.

Different species as well as varieties exhibit significant variations with respect to storage behavior. Different storage methods which have been suggested for the storage of flowers are Normal Refrigerated Storage, Controlled Atmosphere (CA) and Low Pressure Storage (LPS).

Important factors which determine the poststorage life of the flowers are quality of the cut stem to be stored, stage of harvest of the flower, temperature and relative humidity during storage, light, sensitivity of the flower to ethylene and pathogens.

The flowers meant for storage to be stored should be of high quality, free from physical injury and incidence of pests or diseases. Besides, the flowers must also be stored at an optimum harvest maturity. Some flowers like gladioli, roses, lilies, narcissus and irises can be stored at bud stage as the buds of these flowers possess ability to open after storage or transport. The flowers stored at bud stage also last longer than those stored at open stage.

Temperature management during storage is of utmost importance to preserve the quality of the cut flowers. Respiration rates of rose and carnation flowers placed at 20 °C temperature have been found to be nearly 20 times higher than those placed at 0 °C. High respiration rates deplete the flowers of their food reserve due to overutilization of stored carbohydrates. High temperatures also increase water loss. Low temperatures on the other hand, inhibit opening of the buds, ethylene production and growth of the pathogens.

Temperature in the cold room should be precisely controlled because temperature fluctuations lead to condensation of water on the stems which invite pathogens, especially *Botrytis cinerea*. Relative humidity inside the cold room should preferably be kept high (90-95 per cent) to decrease water loss from the stems and prevent desiccation. Only disease-free stems should be stored because high relative humidity in storage is highly favorable for growth of *Botrytis cinerea.* Cold rooms should be kept clean to prevent the spread of pathogens as well as accumulation of ethylene by the decaying plant tissues and microbes.

5.6.1 Refrigerated Storage

Normal refrigerated storage is the simplest and most frequently used method for the storage of cut flowers. It includes both wet and dry storage. In wet storage, the stems are stored with their bases dipping in water or solution of floral preservative, usually a biocidal compound to check the growth of microbes on the stem surface as well as in the vase water. Since during wet storage, the stems continue to absorb water/chemical solution, they exhibit higher metabolic rate as a consequence of which bud development and senescence processes occur more rapidly than in dry storage. Carnation stems stored in water respired 25-30 per cent faster than those stored dry. Wet storage is used to hold the cut flowers for short duration and for their day to day handling. The temperature in wet storage is kept at 2-4 °C *i.e.* slightly higher than that in dry storage. Flowers like iris, lilies and gerbera are suitable for wet refrigerated storage.

In dry storage, the flowers are sealed in moisture-retentive plastic sleeves to prevent the loss of moisture. Dry storage is also referred to as Modified Atmosphere (MA) storage of because the level of O_2 and CO_2 in the storage packages gets modified *i.e.* the concentration of CO_2 increases and that of O_2 decreases due to respiratory activity of the tissues. The suitability of MA packages for storage depends upon the permeability properties of the plastic membrane for CO_2 and O_2. The plastic packages also involve hazards due to accumulation of ethylene, excessive increase in CO_2 or decrease in O_2 levels which create anaerobic conditions. In general, 5 to 7 per cent of CO_2 and 1 to 2 per cent O_2 is suitable for storage of flowers. Build up of very high levels of CO_2 (beyond 4-5 per cent) or extremely low levels of O_2 (lower than 0.4 per cent) in the packages may damage the flowers. Therefore, it is very important to identify suitable films for optimum retention of both CO_2 and O_2 levels inside the packages. Dry storage is more laborious but can be used to hold the flowers for longer duration. Proper precooling of cut stems prior to storage may reduce the risk of damage to flowers caused by high accumulation of CO_2 or ethylene. MA storage of flowers is significant as flowers in these packages lose less water and also occupy less space in storage as well as in transit. Storage in MA packages also holds considerable scope for long term shipment or surface transport of cut flowers. Pulsing of flowers before storage significantly improves their post storage vase life. Pulsing of gladiolus spikes with a solution of sucrose, 20 per cent in combination with aluminium sulphate, 400 mg/l has been found to be very beneficial in extending the vase life of gladiolus spikes after storage in polypropylene packages or after wet storage. Addition of GA_3 to pulsing solution improved the post storage vase life of the spikes. It was reported that pulsing of gladiolus spikes with $AgNO_3$ (1000 ppm) for one hour followed by treatment with 20 per cent sucrose for 24 h before dry storage increased their vase life. Post storage treatment with vase solution containing sucrose (1.5 per cent) plus aluminium sulphate, 300 ppm also significantly improved vase life of cut rose flowers. Spraying the foliage with BA+GA_3 (25 mg/l each) before or after dry storage significantly increased vase life of flowers of Asiatic and Oriental lily. In Anthurium, $AgNO_3$ pulsing for 40 min before dry storage improved vase life of cut Anthurium flowers.

Most flowers are prone to freezing injury below -2 °C but since free water freezes at 0 °C, water droplets condensing on the petals can also cause injury to the tissues. Therefore, the flowers should be stored at 0.5-1 °C. The temperature range for tropical flowers like anthurium, *Cattleya* and *Poinsettia* is 10-15 °C and for sub- tropical flowers such as gladiolus, strelitzia and anemone is 2-8 °C. Response of some important cut flowers to dry and wet refrigerated storage is given in Table 4.

5.6.2 Controlled Atmosphere Storage (CAS)

Controlled atmosphere storage (CAS) refers to low temperature storage in gas tight chambers under decreased levels of O_2 and increased levels of CO_2. CO_2 levels higher than 4-5 percent may produce anaerobic conditions. Major limitation in CA storage of flowers is that the optimum levels of O_2 and CO_2 required for storage may vary for different flowers and hence, different flowers cannot be stored at the same time in the same chamber.

Table 4: Storage Response of some Important Cut Flowers

Storage	*Crop*	*Storage Temperature*	*Maximal Storage Period*
Dry	Carnation	0 to 1 °C	16 to 24 weeks
	Chrysanthemum	1 °C	3 weeks
		4 °C	15 days
	Freesia	0-0.5 °C	I week
	Gerbera	2 °C	2 days
	Gladiolus	4 °C	2 to 3 weeks
	Rose	0.5 to 2 °C	2 weeks
Wet	Alstroemeria	4 °C	
	Anthurium	13 °C	2 to 4 weeks
	Carnation	4 °C	4 weeks
	Chrysanthemum	4 °C	2 weeks
	Dendrobium	5 to 7 °C	10 to 14 weeks
	Gerbera	4 °C	1 weeks
	Gladiolus	4 °C	9-12 days
	Phalenopsis spp.	7-10 °C	2 weeks
	Tuberose	7 to 10 °C	3 to 5 days
	Rose	2 to 5 °C	5 days
	Zantedeschia spp.	4 °C	1 week
	Zinnia	4 °C	1 week

5.6.3 Hypobaric or Low Pressure Storage (LPS)

Hypobaric or low Pressure storage (LPS) refers to storage at low atmosphere pressure *i.e.* 40-60mm Hg under continuous ventilation and high relative humidity. Major disadvantage of LPS is that it causes rapid loss of water and faster removal of gases from the tissues. LPS, however, holds considerable promise for prolonged storage of flowers but due to high cost of installation, it is of limited use.

The optimum storage temperature and duration of storage varies with each individual flower. In general, temperate flowers *e.g.* rose, carnation, lily, cymbidium and chrysanthemum are best stored at a temperature close to 0 °C *i.e.* above the freezing point of plant tissues. Flowers originating from tropical regions *e.g.* anthurium, vanda, cattleya and bird of paradise are sensitive to chilling injury and hence, have to be stored at higher temperatures.

Flowers exhibit decline in vase life and the buds also lose the ability to open after prolonged storage . Though precise mechanism underlying decrease in vase life of flowers following storage is not clearly understood, loss of carbohydrates, accumulation of ethylene in the atmosphere surrounding the flowers, loss of membrane phospholipids, changes in activity of membrane enzymes, transport systems and hormone binding systems as well asaccumulation of toxic compounds like acetaldehyde and ethanol have been suggested to affect poststorage vase life of flowers.

Many leafy stems *viz.* alstroemeria, asparagus, miniature gladiolus, lilium and chrysanthemum are prone to poststorage yellowing of leaves. Dark conditions during prolonged storage hasten the yellowing of leaves in these species. The longevity of leaves can, however, be improved by the application of plant growth regulators like cytokinins, gibberellins though the efficacy of these growth regulators differs with the species. The effective treatments which can be employed to delay yellowing of leaves in some important species are presented in Table 5.

Table 5: Suitable Treatments to Prevent Premature Yellowing of Leaves in some Important Flower Species

Name of the Flower	*Treatment*
Alstroemeria pelegrina L.	☆ GA_3, 10^{-5} M
	☆ TDZ, 10mM for 24 h
Chrysanthemum	☆ Methanol, 4 per cent continuous or BAP, 10 ppm as 24 h pulse
	☆ TDZ, 10-100 mM for 24 h
	☆ BAP, 100 ppm spray
Easter lily (*Lilium longi florum* Thunb.	Spray with GA_3+BA (500mg/l each)
Stock (*Mathiolain cana* L.)	TDZ, 5mM for 24 h
Tulip (*Tulipages neriana)*	TDZ, 10-100 mM for 24 h depending upon the variety

5.7 Postharvest Disorders and Diseases

Cut stems of some ornamental species are prone to physiological disorders whichcause decline in their longevity and rejection in the market. These disorders include bent neck of cut roses, bluing and blackening of petals in red coloured cultivars of rose, topple of tulips, stem break in gerbera, yellowing and blackening of leaves and geotropic bending of stems. These are discussed below.

a) Bent Neck of Cut Roses

Bent neck of rose is a physiological disorder wherein the stem fails to hold the bud erect which droops in the peduncle region. Bent neck usually occurs in the stems which are harvested when the bud is immature or when cut stem is under severe water stress conditions. The bent neck in stems cut at too immature stage occurs due to lack of sufficient hardening (lignifications) in the peduncle region. The leafy stems of rose also lose considerable quantities of water through transpiration and are hence, prone to water stress conditions. Under severe water stress conditions, buds draw water from the peduncle which lies immediately below it. This leads to loss of turgidity in the peduncle cells which become flaccid. The incidence of bent neck also varies in different cultivars. Some cultivars like Cara Mia are highly susceptible to bent neck. In order to prevent bent neck, the stems should be harvested in the morning when they are fully turgid and at proper harvest maturity when the peduncle region is sufficiently lignified. The stems should not be placed out of water for prolonged duration. The wilted stems after prolonged storage or transportation should be placed neck deep in tepid water so that they regain turgidity.

b) Bluing and Blackening of Petals in Rose

Petal bluing in red varieties of rose occurs due to carbohydrate starvation and subsequent hydrolysis of proteins and respiratory loss of amino acids which lead to increase in free ammonia in the cells. Increased ammonia in the cells increases cytosolic pH which causes bluing of petals. Bluing of petals occurs to copigment complex of cyaniding 3,5 diglucoside, quercetin and kampferol glycosides. Treatment with preservatives containing sugars can be used to prevent bluing of petals. Petal blackening also occurs in some red cultivars of rose due to calcium and perhaps boron nutrition.

c) Stem Break of Gerbera

Breaking of the stem in the middle portion is commonly observed in many cultivars of gerbera. Stem break occurs due to development of water stress conditions in the stem. The cut stem of gerbera absorbs water through two different pathways: (a) direct one through the xylem vessels at the cut surface and (b) an indirect one through the cavity in the stem. Stem break occurs when the direct water uptake is inhibited by bacterial activity and the indirect uptake is hampered. The water imbalance caused by bacterial growth in xylem vessels is the major cause of stem break in gerbera. Pretreatment of the stems with one per cent sodium hypochlorite (NaOCl) for one min in combination with wetting agent such as Tween-20 (0.1 per cent) has been recommended to prevent stem break in gerbera. Likewise, impregnation of the stem with 1200 ppm $AgNO_3$ for 10 min or 20 ppm $AgNO_3$ in vase solution has also been found to completely inhibit microbial growth in vase water.

d) Topple of Tulips

Topple of tulips is associated with calcium deficiency and high atmospheric humidity. The stem collapses a few cm below the base of the flowers at the time of flowering or picking. The scape, generally at an upper internode becomes dark green and watery. The affected areas shrink and the scape collapses and topples under the weight of the flower. Under severe calcium deficiency, the symptoms also appear on the scape or on the leaves. Sprays of 2 per cent solution of calcium nitrate solutions can reduce the losses.

e) Yellowing and Blackening of Leaves

In general, the leaves last longer than the flowers. However, in certain species, leaves show yellowing or blackening much before senescence of the flower bud. Yellowing and blackening of leaves marks the end of display life in some flowers.

Yellowing of leaves is a major problem in certain cultivars of alstroemeria, chrysanthemum, statice, lilies etc. and occurs due to breakdown of chlorophyll promoted by high temperatures as well as dark conditions during storage. Yellowing of leaves can be delayed by the application of gibberellins, kinetin, benzyl adenine and thidiauron. The response of growth regulators, however, varies with the species.

Blackening of leaves is of common occurrence in certain species of *Protea* and results from polymerization of phenols and tannins. It is mainly induced by water stress, high temperature and dark conditions and competition for carbohydrates.

Vase solutions containing sugar helps to reduce this order. The concentration of sucrose higher than 0.5–1 per cent may, however, accelerate blackening, since proteas are also sensitive to exogenously-applied sugar.

f) Geotropic Bending

In certain flowers like gladiolus, antirrhinum, larkspur, lisianthus, stock, lupine etc. the flowering stems are sensitive to geotropic bending *i.e.* if placed horizontally, sub-terminal portion of the stem shows upward bending which is referred to as "geotropic bending". Geotropic bending occurs due to cell elongation in the sub-terminal region of the stem due to downward lateral movement of the auxin form the tip region. Geotropic bending leads to loss of quality and the affected stems are not preferred in the market. Low temperature conditions prevent geotropic bending but the stems regain the ability to bend when transferred to normal conditions. In order to prevent geotropic bending, the stems should always be held vertically during storage or transport.

g) Postharvest Diseases

Botrytis blight caused by grey mould fungus (*Botrytis cinerea*) is a major postharvest disease of cut flowers causing substantial damage. The disease develops very rapidly under high moisture and low temperature conditions. The flowers producing nectar may also be more susceptible to grey mould. The pathogen is usually carried along with the cut stem from the greenhouse or field conditions in which the crop is grown. Adequate measures are, therefore, required to be taken to control this disease in the field or greenhouse conditions. Especially, for storage only disease-free flowers or foliage should be used.

5.8 Grading

After cooling, the stems are shifted to air-conditioned grading room. In the grading room, defective stems with blemishes or symptoms of diseases or pest infestation are removed. The stems are sorted out in different grades of length manually or by colour coded board or automatic grading machines. Grading refers to categorization of flowers on the basis of quality parameters. For any market, it is important that each bunch is made up to the same size, weight or quality of the bloom. The bud should be at right harvest maturity, free from blemishes or injuries due to attack of pathogens, insects or pests, straight and strong enough to stand the weight of the bud. The bud should be bright in colour. The foliage should be dark green and free from the dust or residue of fungicide or pesticide. The stem should be free from physiological disorder such as bent back (in roses), tip bending (in gladiolus and other flowers), stem break (in gerbera), calyx splitting (in carnation) etc. The Economic Commission for Europe (ECE) based at Geneva and the Society for American Florists have recommended grades for some species of cut flowers but at present, there are no uniform grading standards for cut flowers in the world. However, the growers can get guidelines from the above-mentioned categorization and accordingly, grade the flowers. It is, however, very important for the growers to first collect information from the flower market where they intend to market produce and judge for themselves as to what type of flowers fetch remunerative price in the market.

Under the grading system suggested by the Society of American Florists, the flowers are categorized as blue (Fancy), red (Special), green (Standard) and yellow (Utility). The blue (Fancy) grade includes the best quality flowers. The U.S. standards for hybrid tea rose, gladiolus, carnation (standard) and chrysanthemum (standard) are given in Table 6.

Table 6: Standards for some Cut Flowers by the Society of American Florists

Gladiolus

Grade	*Spike Length (cm)*	*Minimum Number of Florets*
Blue (fancy)	107 and above	16
Red (special)	96-107	14
Green (standard)	81-96	12
Yellow (utility)	81 and above	10

Carnation (Standard)

Grade	*Minimum Flower Diameter (mm)*	*Minimum Length Overall (cm)*	
Blue (Fancy)	Tight	50	
	Fairly tight	62	55
	Open	75	
Red (Standard)	Tight	44	
	Fairly tight	56	43
	Open	69	
Green (Short)	Tight	none	
	Fairly tight	none	30

Chrysanthemum

Grade	*Minimum Flower Diameter (mm)*	*Minimum Length Overall (cm)*
Blue	140	76
Red	121	76
Green	102	61

Rose

Grade	*Minimum Length Overall (cm)*	
	Hybrid Teas	*Sweet Hearts*
Blue	56	36
Red	36	25
Green	25	15

Leaves from the lower 4-5 cm portion of the stem should be removed and stems put in buckets containing chlorine (50 ppm), aluminium sulphate or citric acid (300 ppm). These chemicals disinfect the stem and check microbial growth which otherwise could reduce shelf life of the stems. The graded stems are immediately shifted to cold rooms till packed.

5.9 Packaging

The flower should be graded and bunched before treating with chemicals or before storage. Flowers of orchids, standard chrysanthemum, anthurium, gerbera and some specialty flowers are, however, packed individually. The number of flowers in a bunch varies with the species and requirement of the market. Flowers are generally grouped in to bunches of 10,12 or 20 stems, and loosely tied with rubber band to secure the buds in position. The bunches should be carefully put together to prevent any damage to the blooms and leaves. Individual flowers of gerbera, orchids, anthurium and standard chrysanthemum which are harvested at open stage, are packed individually so as to prevent any likely damage to the open buds due to friction. The individual flowers are protected by paper, cellophane or polyethylene sleeves. Cushioning material such as shredded paper is also used to further reduce damage due to friction. The bases of tropical flowers like dendrobium and anthurium are kept moist by placing them in specially designed vials filled with water or in moist wool. The bunches are held in polyethylene or cellophane sleeves or buds as in roses are wrapped in corrugated paper to protect them from mechanical damage. Different flowers require different methods of wrapping or sleeving. For instance, the buds of rose are wrapped with 2 ply soft corrugated paper. The wrapping paper should project at least 2-3 cm above the bunch to protect the buds from thorns. Long-stemmed varieties such as Grand Gala or First Red have bigger-sized buds. These buds can be packed in 5 different layers of 4 flowers in each layer at two different levels. Such staggered bunching is useful to avoid damage to the buds by mutual pressure. Short-stemmed varieties have small-sized buds which can be packed keeping all the buds in the same plane. The bunches are packed in precooled 4-5-ply teleboxes (top completely overlapping the bottom) of fibreboard in such a way that lower heads face in opposite direction *i.e.* towards the width. The heads of the flowers should be placed 5-10 cm from the end of the box to allow effective precooling and prevent damage due to bruising. The bundles should be placed in rows along the width. The second layer is placed opposite to the first layer for proper utilization of the space. In the packing box, the layers of flowers are covered with soft paper. The quality board (usually 4-5 ply boxes should be strong enough to hold the weight of at least eight such completely filled boxes) should be used for packing the flowers. Some growers use newspaper to cover the stems but care should be taken to ensure that the newsprint does not leave marks on the stems. After packing the flowers, the box is closed with the lid and strapped properly. Mixing of varieties or stems of varying lengths in the same box should be avoided. The stems should be tightly held inside the boxes so that they are not subjected to jerk movements during transport. The boxes should be properly marked with symbol 'Fresh Flowers' on both sides so that they are handled with care at the airports. Besides, the boxes should have stickers indicating name of the variety, number and length of the stems it contains. For international market, the flowers of gerbera are packed in flat boxes containing paper inserts with holes for individual stems. The support is necessary to keep the stems in place. For domestic markets, however, before packing the individual flowers are inserted in polyethylene sleeves to protect them from bruising.

Polyethylene sleeves are generally employed for wrapping a wide range of flowers but such a wrap may tend to retain moisture and encourage fungal attack. Hence, many wholesalers prefer naked bunches or perforated polythene sleeves. Hot-needle sleeves which have small, perforated holes allow the bunches to breath and reduce the chances of fungal attack. The flowers are held tightly in the boxes so that they are not damaged by the jerk movements during transit. The dimensions of the packing boxes are worked out keeping in view the length of the stem, type of the flower to be packed and efficient utilization of the space in the cargo or in refrigerated trucks.

Over packing and under packing both cause excessive damage. A wooden batten should be placed across the centre of the box under light pressure and nailed to the sides to prevent the movements of bunches during transport. The dimension of the packing boxes, however, varies from grower to grower and depends upon the length of the stem and number of stems to be packed as well as type of the cargo space available in the aircraft so that the boxes can fit well into the available spaces.

5.10 Transportation

Flowers such as gladioli, snapdragon and lupine which show geotropic bending when placed in horizontal condition during storage or transport should be transported in vertical hampers. A new packaging system, the 'Procona' system developed by Pagtar Innovations, Roosendaal, The Netherlands allows transport of flowers upright in water and is very useful to keep flowers hydrated and fresh during transport over long distances, especially when they are not transported under proper temperature conditions. 'Procona System' is a package made up of plastic container, a cardboard collar that fits the container and a lid with many ventilation holes. The corners of the lid are very strong so that the packages can be stacked without getting damaged. The lid also has ridges so that the packages do not shift when stacked on the lid. When stacked, the container does not shut off the lid of the package below. As a consequence, the water vapours can escape through the ventilation holes in the lid. During transportation the flower heads receive fresh, cold air through the ventilation holes. The consumer can also see the colour of the flowers through the ventilation holes in the lid. Since the flowers stand in water during transportation, they maintain their quality.

Flowers are usually lifted by air and by refrigerated vans/trucks. But when the distances are short, non-refrigerated insulated trucks are also put to use for transportation. Temperature management during transit is essential.

Therefore, the flowers and the boxes are essentially required to be cooled before transport. Once packed, the flowers are difficult to cool.

The simplest method to cool the flowers is to pack them in the cold room. This method is, however, labour intensive and takes longer to cool the flowers. Forced air cooling of flowers in boxes with closeable side holes is the most common and effective method for cooling of flowers. Cold air is sucked or forced into the boxes to reduce their temperature quickly. Forced air cooling is efficient and the flowers can be cooled in about 45 min to one hour.

Vacuum cooling is also the quick and cost-efficient cooling system for cut flowers. The technology is based on the principle that water boils at lower temperatures as the pressure decreases. At about 1/10th atmospheric pressure, water boils at 0 °C. In the vacuum cooler, the pressure is reduced to a level where water starts boiling at 1-2 °C. The boiling process removes heat from the product which can be cooled down to 1 or 2 °C within 15-45 min depending upon the type of the product and amount of heat required to be removed. Although this system cools leaves or stems efficiently, flower petals often take longer to cool as water does not escape readily from them.

Optimum temperature to which the flowers are cooled varies in different flower species. Flowers are generally cooled at 0-1 °C. Care should be taken to pack the flowers in such a way that cold air flows freely in the boxes. If the flowers are to be transported through refrigerated vans, the side holes should be kept open to remove heat generated through respiration and free circulation of cold air inside the boxes. The flowers originating in tropical and sub-tropical climates are prone to chilling injury and are required to be maintained at higher temperatures (as discussed under "Storage of Flowers").Before loading it is to be ensured that the truck or van is precooled to the ideal temperature. Before loading all boxes are also required to be precooled to the optimum transportation temperature. To avoid the box becoming warm, loading is to be carried out as quickly as possible.

6.0 Domestic and Foreign Trade

Flowers from India are marketed both within India and are also exported to far places such as Europe and Japan by air transport. The boxes are taken to the airport in the refrigerated vans to keep the flowers cool. The international airports also provide cooling facilities to the growers. In case of limited space in the cold room, air conditioning unit of the refrigerated truck can be plugged to the electric points provided by airport authorities to use refrigerated trucks as a makeshift cold store till the flowers are loaded in the aircraft. It should be ensured that the flowers are loaded in a special cabin meant for transport of perishable commodities.

In the international flower trade, the flowers are traded through auction markets as well as directly through whole sale and retail outlets. In our country, the trade in cut flowers is rather of recent origin and hence, there is no organized marketing system of flowers. Mostly, the flowers are marketed by the growers directly to the retail outlets. Some efforts have been made in some states to develop auction centres for flowers.

Karnataka Agro Industries Corporation (KAIC) has started a flower auction centre at Hebbal, Bangalore which possesses state-of-the-art electronic auctioning system and automatic cold storage system with a capacity to store 2.5 lakh stems. At present, the centre deals more with domestic trade but will be converted into an international auction centre. Maharashtra Agro Industries Corporation has also established a flower auction centre at Mumbai, which is yet to become operational. Similarly, flower market at Gaziabad, Uttar Pradesh is also operational. Ghazipur Flower Trade Centre, Ghazipur on the outskirts Delhi gets hundreds of varieties of flowers from various parts of the country and even from some south Asian countries like Thailand, Indonesia, etc. In the international market, flowers are auctioned in the auction centres. There are seven auction centres in Holland. The largest flower auction centre in the world is in Aalsmeer. The flowers from different countries of the world are traded at these centres. Nearly 60 per cent flowers are traded in Aalsmeer and Naaldwijk/Bleiswijk auction centres.

7.0 Conclusion

By observing the guidelines mentioned below, vase life of the flowers can be significantly improved. 1. Flowers should always be harvested at an appropriate stage of development. Too tight buds fail to open whereas mature flowers do not last long. 2. After harvesting, the stems should not be placed out of water for more than a few minutes. 3. The vases should be disinfected with chlorinated disinfectants before putting the flowering stems. The unclean vases harbor many microbes which multiply very rapidly when the stem is placed in the vase. 4. Water in the vase should be changed at 2-3 days interval to prevent growth of microbes in vase water. 5. About 2-3 cm basal portion of the stem should be recut at 2-3 day intervals to expose fresh tissues for improving water uptake. 6. Lower leaves dipping in water/solution should be removed immediately as these leaves emit foul smell on rotting and also promote microbial growth which decreases vase life of the stem. 7. Vases should be placed at cool places. 8. Withered flowers should be periodically removed from the stem as they release considerable quantities of ethylene and cause senescence of other flowers.

8.0 Important References

Bala, M., Kumar, R. and Singh, K. 2008a. Effect of pulsing and holding solutions on keeping quality of cut flowers of rose. *J. Orna. Hort.* 11(1): 54-57.

Cho, M., Celikel, F.G., Dodge, L. and reid, M.S. 2001. Sucrose enhances the postharvest quality of cut flowers of *Eustomagrandiflorum* (Raf.) Shinn. *Acta Hort.* 543:

Han, S.S. 2001. Benzyladenine and gibberellins improve postharvest quality of cut Asiatic and Oriental lilies. *HortScience*, **36**: 741-45.

Karki, K., Kumar, S., Srivastava, R. and Jauhari, S. 2004. Effect of floral preservatives on vase lfie of cut roses grown underpolyhouse conditions. *J.Orna. Hort.* 7: 121-23.

Meeteren, U. van. 1978. Water relations and keeping quality of cut gerbera flowers. I. The cause of stem break. *Scientia Hort.* **8**: 65-74.

Mor., Y., Reid, M.S. and Kofranek, A.M. 1984. Pulse treatment with silver thiosulphate and sucrose improve he vase life of sweet peas. *J. Amer. Soc. Hort. Sci.* **109**: 866-68.

Namita, Kumar, R. and Singh, K. 2006b. Response of vase solutions onkeeping quality of cut spikes of gladiolus. *J. Orna. Hort.* **9**: 296-97.

Nilofar and Paul, T.M. 2008. Postharvest management of gladiolus. *J. Orna. Hort.* **11**: 69-71.

Pal, A., Kumar, S. and Srivastava, R. 2003. Effect of floral preservatives on postharvest management in gladiolus spikes. *J. Orna. Hort.* 6: 367-71.

Reid, M.S. and Kofranek, A.M. 1980. Postharvest physiology of cut flowers. *Chronica Hort.* **20**: 25-27.

Serek, M., Sisler, E.C. and Reid, M.S. 1995a. Effects of 1-MCP on the vase life and ethylene response of cut flowers. *Plant Growth Regul.* **16**: 97-98.

Singh, K., Arora, J.S. and Bhattacharjee, S.K. 2001a. *Postharvest Management of Cut Flowers.* Tech. Bull. No. 10, All India Coordinated Research Project on Floriculture, IARI., New Delhi, pp. 39.

Singh, K., Singh, P. and Kapoor, M. 2007. Effect of vase and pulsing solutiuons on keeping quality of standard carnation (*Dianthus caryophyllus* Linn.) cut flowers. *J. Orna. Hort.* **10**: 20-24.

Singh, K., Singh, P.J. and Arora, J.S. 2003. Effect of sucrose on vase life of cut roses. *J. Orna. Hort.* **6**: 145-46.

Singh, K., Singh, P.J. and Kumar, R. 2004a. Effect of chemicals on keeping quality of cut roses. *Adv. Hort. Sci.* 18. 161-67.

Chapter 30
Garlic

K.E. Lawande

Vice Chancellor, Dr. Balasaheb Sawant Konkan Krishi Vidyapeeth, Tal. Dapoli, Dist. Ratnagiri – 415 712, Maharashtra, India

P.C. Tripathi

Head, Central Horticultural Experiment Station (IIHR), Chettalli – 571 248, Karnataka, India

E-mail: kelawande2011@gmail.com;
E-mail: prakaashtripathii2000@yahoo.co.in

1.0 Introduction

Garlic (*Allium sativum* Linn.) is used around the world for its pungent flavour as a seasoning or condiment. It is a major constituent of many ayurvedic medicines. Animal studies, and some early research studies in humans, have suggested possible cardiovascular benefits of garlic. Many studies found garlic supplementation reduced accumulation of cholesterol on the vascular walls of animals and in humans. Garlic is produced only in rabi season in plains. It comes from MP in February and continues up to March-April. In Southern hills particularly Nilgiri hills garlic is cultivated in two seasons *i.e.* March-April to July-August and sometimes also in October to March-April. In Northern hills it is cultivated in one season *i.e.* October to June.

2.0 Status of Cultivation

India ranks second in the area and production of garlic in the world after China. In India garlic is grown over 2.45 lakh hectare area with the production of 12.25 lakh tonnes. Gujarat and Madhya Pradesh are the leading states in India contributing more than 45 per cent of both area and production. The other major garlic growing states are Rajasthan, Uttar Pradesh and Maharashtra. Garlic is a traditional exportable crop of India, however, export from the India is not stable.Garlic export form India is largely influenced by the production and the domestic demands of the country. Currently small clove garlic is exported for consumption of ethnic Indian population in Middle East and South East Asian countries *viz.*, Bangladesh, Indonesia, Philippines and Gulf countries. A large quantity of big clove garlic is also imported from China and other countries when domestic production is low.

3.0 Major Commercial Varieties

More than 20 varieties of garlic have been released from different research organization. The varieties Godavari and Sweta from at MPKV, Rahuri, HG-1 and HG-2 at HAU, Haryana, Pusa Selection from IARI, New Delhi, LCG-1 from Ludhiana, ARU-52, Almora, VL-6 and VL-7 from VPKAS, Almora, Agrifound white, Yamuna safed-1, 2,3, 4, 5 and Agrifound Pravati, Agrifound Parvarti-2 from NHRDF, Nasik, Bhima Omkar, Bhima Purple, DOGR, Rajgurunagar were released during last 30 years. Most of these varieties have smaller bulbs and numbers of cloves per bulb are more than 20 cloves. Some small clove varieties released have purple colour while others are white. These varieties are suitable for export purpose for those countries which demand small clove garlic. The export of selected bigger size bulbs (more than 40 mm) of small cloved landraces is done from Madhya Pradesh. There are few bigger cloved cultivars released for cultivation. Yamuna Safed-3 (G-282) and Yamuna Safed-5 are released for plains and AgrifoundParvati (G-313), AgrifoundParvati -2 are released for Northern hills. Big clove garlic varieties are cultivated in Uttarakhand, Himachal Pradesh and Jammu and Kashmir. Some these varieties can also be cultivatedin Kodaikanal and Mettupalayam in Tamilnadu. The bulbs of these varieties are white and 5.0 -6.0 cm in diameter and numbers of cloves per bulbs are less than 20 and have good shelf life. These are suitable for export.

4.0 Nature and Causes of Postharvest Losses

During the period of storage, 30-40 per cent losses occur mainly due to loss of moisture, sprouting and rotting. More than 95 per cent produce of garlic arrives in market in the months of March to May. But there is constant demand of garlic in the market as it is required daily in small quantity in almost all houses. Further, there is glut in the market during the harvest season. As a result farmers do not get the appropriate prices and they are forced to sale their produce at lower price. Thus it becomes essential to store garlic for regular supply in the market and to stabilize the price. It is estimated that 8-9 lakhs tonnes of garlic need to be stored to fulfil the domestic and export demand. Generally manage-ment of storage losses is initiated after the harvest of the crops. There are many pre harvest aspects which affect the storage of garlic. These includes genetic makeup of cultivars, type of manure and fertilizer and their time of application and quality,quantity, method and time of application of water, pre harvest treatments, harvesting time etc. Postharvest factors include field curing, drying, store environment, packaging, transport etc.

5.0 Postharvest Management

5.1 Harvest Maturity

The change of leaf colour is the sign of maturity in garlic. The stage of maturity of cloves at harvest has significant influence on dormancy.As the bulbs reach maturity, the plants cease to produce new leaves and senescence starts. The tops just above the neck of the bulbs weaken and bend down but still remain green. Further, there is development of bulbils in the pseudostem just above the ground. This is also an indication of maturity in tropical type of garlic. Bulbs continue to increase in size and weight for sometime after softening of the neck fall. This increase is due to continued accumulation of dry matter in the bulb. Bulbs harvested after neck fall have been reported to take less time for drying and curing. Garlic harvested when the leaves are still green, require more time to sprout than mature bulbs harvested when the foliage has dried. It has also been reported that harvesting of garlic bulbs before the tops have fallen over indicating full maturity, results in poor storage whereas delayed harvesting results in splitting of bulbs and re-sprouting of bulbs.

5.2 Harvesting

Harvesting is done manually by uprooting. Some time the hand hoe is used in hard soil for harvesting of garlic but it increased the chances of damage of the bulbs. These bulbs get disease infection and exhibits more losses in storage. After harvesting the bulbs are is left in the field for 3-4 days for removal of excess surface moisture and drying of leaves. This allows the formation of strong outer protective scales and hardening of neck. Garlic bulbs, after harvest, are cured naturally in the field by windrowing in such a way that the bulbs of one row are protected from the sun by the foliage of the adjacent row. After field curing, the bulbs are cured in a ventilated shed for 7 to 15 days either with tops or after cutting the tops leaving about 2.5 cm neck intact and roots are removed completely. These bulbs are than graded as per size.

5.3 Curing

It is extremely essential that the garlic bulbs destined for storage should be well cured in the field. The garlic bulbs should be provided with adequate air circulation through the storage containers to remove the transpired moisture. Incomplete curing results in rapid decay particularly when the bulbs are held above 0 °C.

5.4 Sorting and Grading

After removal of leaves the damaged, sprouted and diseased bulbs are removed. Then the bulbs are graded as per size. The size of bulbs varies in different varieties or land races. The grading is mostly done manually but now size graders are available in India. The details of grade designation and definition of different qualities of garlic are mentioned in Table 1.

5.4.1 General Characteristics

i) The bulbs shall be reasonably of uniform in colour, shape and pungency characteristics of the variety type.

ii) The bulbs shall be well cured and dried.

iii) Bulbs shall be compact with cloves well filled and fairly plump.

iv) Bulbs shall be free from damage caused by moulds, decay, splitted cloves, dirt, sun burn, dust, sprouting etc.

Table 1: Grade Designations and Definitions of Quality of Garlic

Grade	*Size (Minimum diameter in mm)*			
	Pune/Nasik Garlic	*Madurai Plain Garlic*	*Nilgiri Hill Garlic*	*Jamnagar Garlic*
Big	30	20	35	35
Medium	25	15	25	25
small	10	10	10	10

The tolerance for size is such that incidental error in sizing is not more than 5 per cent by weight of the next lower grade. The tolerance for requirement in respect of general characteristics is allowed for variations other than size. Incidental error due to improper grading and handling not more than 10 per cent by weight of garlic in any containers may be below the requirement specified under general characteristics but not more than a total of 1/5th of this tolerance or 2 per cent shall be allowed for garlic which is affected by decay.

5.5. Packaging

5.5.1 Domestic Market

Garlic is packed in Hessian cloth bags/jute bags of varying capacity for domestic market. Recently nylon net bags or leno bags are used for domestic market. Garlic is packed in 90kg packing in Andhra Pradesh. But generally 40-60 kg capacity jute bags are used for domestic marketing in Karnataka, Maharashtra,Gujarat, Madhya Pradesh and other garlic growing states. Packaging in small size bags 5 to 25 kg capacity is also gaining popularity in domestic market.

5.5.2 Export Market

Garlic grading and packing rules provide for packing size of 18 kg, 25 kg and 50 kg. For 50 kg packaging jute bags are preferred and 18 kg and 25 kg packing is done in perforated corrugated cardboard boxes for export. The packages are labelled with information about variety, weight, the address of supplier and other details. The strength of the jute bags used is less. Due to this there is loss during various process of handling many times. Preference of plastic woven bags is increasing due to their attractiveness and easy visibility of the quality of garlic in the bags. Although in Europe and other Western countries preference is given for jute bags because of disposal problem in plastic woven bags.

5.6 Storage

Garlic is stored for about 7 to 8 months with periodical release for domestic and export purposes. Since garlic is produced mostly in March and April, it has to be stored for augmenting continuous supply to the market. The storage of garlic besides genotypic factors and pre harvest management, largely depends on storage conditions. There are two distinct temperature regimes where storage losses are minimum. One is high temperature regime, where storage temperature is 25-30 °C and another is low temperature regime, where temperature is augmented maintained 0 to 2 °C. Best results under both the temperature regimes are obtained when humidity is maintained at 65 to 70 per cent. The storage losses in high temperature conditions (25-30 °C) are high (20-30 per cent) but storage cost is low, while in low temperature conditions (0-2 °C) losses are minimum (0.5 per cent) and storage period is longer. However, storage cost is high. High temperature storage (more than 30 °C) at ambient leads to high weight loss while storage at lower temperature (less than 10 °C) enhances sprouting losses. High humidity (more than 70 per cent) coupled with high temperature enhances storage diseases, while low humidity enhances weight loss in storage.

Garlic should be stored with leaves hung by their tops or stored in circular heaps show minimum losses at high temperature regime. This practice is adopted for garlic storage in most of the garlic growing areas of the country particularly for domestic consumption and seed purpose.The storage of garlic with dried foliage in bundles tightly heaped in ventilated storage structures is done at NRCOG and also by farmers in Nasik/Pune area which gives low storage loses. However, the presence of leaves for marketing and large scale storage is undesirables as it affects the grading, packing and marketing. Thus the leaves of garlic are removed after curing or sometime after on farm storage. The garlic storage structures should be designed in such a way that the optimum conditions of temperature and moisture maintained. In India, cold storage of garlic is not done commercially because of several reasons and almost all garlic is stored at room temperature conditions. Various types of storage structures have been designed and tested for garlic storage. Among these structures, bottom and side ventilation of 10, 25, and 50 tons capacity are recommended. Low volume structures are with single row, while high volume are with double row and a gangway between two rows. Studies on width and height of heap revealed that 120 cm staking width and 150 cm of height is most appropriate. Any increase beyond this recommendation increases storage losses.

Garlic can be stored in cold store at 0-2 °C temperature and 65-70 per cent relative humidity from May to January. The results revealed that total losses in cold storage were significantly lower (11.5 per cent) than at ambient condition (33.9 per cent) after 5 months storage. Physiological weight loss, diseases and drying of cloves was lower in cold store than the at ambient. However, when these cold stored bulbs were taken out and kept at ambient storage conditions, rapid sprouting was observed and it was 13.3 per cent after 20 days and further increased. The post cold storage induced sprouting can be prevented with irradiation of garlic before storage. Irradiation with 2-6 krad of Cobalt 60 gamma rays have also been recommended for controlling sprouting in storage. The irradiation given to the bulbs within 8 weeks of harvest inhibited sprouting effectively.

6.0 Transportation

Harvested garlic is transported to the collection centre in open bullock carts or tractor trolleys. It is also brought in jute bags or nylon netted bags in trucks. While loading bags in trucks, care should be taken in loading upto 6-8 bags height of stacking as over loading will damage the garlic bulbs in the bottom layer of bags by splitting. The split bulbs do not get good rates in market and increase the chances of rot. Garlic bulbs are generally sold to local whole sellers after reaching to collection centres. These transactions usually take place with the help of commission agents. The exporters either buy their requirements directly from collection centres through local commission agents or from whole sellers and then transport to ports by trucks after grading and packing.

7.0 Domestic Marketing and Export

Garlic is a traditional export item. In 1961-62, India exported about 2000 tonnes of garlic. Garlic is exported to Quatar, Saudi Arabia, United Arab Emirates, Bahrain, Mauritius, Kuwait, Bangladesh and Sri Lanka. The export has been ranging from 0.10-3.0 per cent of domestic production. Exporters of garlic purchase their requirement from important assembly markets in Gujarat, Madhya Pradesh, Tamil Nadu, Karnataka and Uttar Pradesh. The export of garlic has also gone up from 2000 tonnes in 1961-62 to about 23,000 tonnes during 2012-13. The export of garlic however, came down to 578 tonnes during 2002-03 and 182 tonnes in 2007-08 due to less availability. The import of garlic was very less till 1997 because of restricted imports of garlic with specific licence. However, Govt. of India has lifted the restrictions. Thereafter import has substantially increased in recent years. Apart from fresh garlic, garlic products are also imported (Table 2).

Table 2: Export and Import of Garlic in India (1998–2014)

Year	*Export (Tonnes)*	*Import (Tonnes)*
1998-99	3978.00	5189
1999-00	8542.00	11636
2000-01	4442.55	2984
2001-02	657.00	36187
2002-03	578.00	40576
2003-04	1567.23	18039
2004-05	2236.47	19534
2005-06	32495.24	–
2006-07	12210.79	–
2007-08	182.47	–
2008-09	858.69	–
2009-10	19767.32	–
2010-11	21827.16	–
2011-12	1325.60	–
2012-13	22665.00	–
2013-14	23,869.27	–

7.1 Export Requirement

In earlier times the export demand for Bangladesh, Sri Lanka and Middle East countries was of small cloved types. However, bigger cloved garlic bulbs are now in demand from all over the world. Preference is for garlic bulbs having 40-60 mm diameter with 10-15 cloves in each bulb. Philippines still require smaller cloved varieties. Govt. of India has prescribed certain grade designation for different qualities of garlic for export.

7.2 Non-availability of Export Worthy Stocks

There is not much organised production in garlic for export. There is no uniformity in size, shape or colour. There is splitting of the bulbs due to delayed harvesting or other factors. Adequate care in harvesting and sorting at farmer's level is not taken. All these lead to availability of poor quality stocks at the market yard. Most of the varieties of garlic were released during last 30 years and still sizable area is under land races and variety wise material is not available for export

7.3 High Domestic Demand and High Ruling Prices

The domestic demand of garlic is high. Garlic crop is mainly produced in those areas which are prone to drought that affects production. The failure of crop in those years cause shortage in supply for domestic market and prices of garlic in domestic market goes up and usually no stock remain left for export. This creates problem in regular supply of garlic for export. Due to this, many times it is not possible to compete with other countries resulting in reduced export.

7.4 Lack of Proper Postharvest Practices and Quality

New exporters in garlic export business are not aware of proper postharvest practices and also quality standards for export. Many exporters do not have adequate facilities for handling the produce and sorting, grading and packing is done with the help of local packers where precautions for proper checking about quality, weight etc. are not taken properly by the packer as well as exporter.Appropriate sorting and grading is not practiced as per the requirement of the importers. This does not only affect export but hampers the image of the country.The stocks of the garlic in many times kept in open sun by traders and exporters which results in sun scalding. Many times farmers bring the stocks without proper curing. If such stocks are packed without removing field heat, the chances of rot increases during transit. The hooks used during loading and unloading on the trucks or in wagons also injure the bulbs.High temperature, high humidity and improper ventilationduring transit increases respiration rate which results in rotting of the bulbs.

7.5 Inadequate Transport Facilities

Transportation from producing centres to port towns is done by rail and road both. Rail transport though is cheaper takes longer time. Sometimes wagons are also not available. Further there is no ventilation in the wagons. Road transport though takes lesser time it is costlier. Adequate trucks at a time are also not available. In case of rail transport sometimes it is difficult to synchronise the movement of the stock by rail with the sailing schedule of vessel. Speedy rail transport with insulated wagons will lessen the cost of transportation.

7.6 Non-availability of Adequate Storage and Handling Facilities

At ports, many times storage is to be arranged before vessel commences loading. Garlic bags are stacked in sheds without having any ventilation and also stacking height is very high. While handling, the labourers sit on the bags or walk on the same. Further there are no suitable vessels for carrying garlic from Indian ports. It has also been observed that even for loading, the vessels have to wait for getting berth like general cargo vessels resulting in inordinate delay in loading. Vessels do not directly go to the final destination but touch many other destinations as a result more time is involved thereby bulbs are damaged. Now containerised shipments are more preferred for door to door delivery. The problem in these containers is this that there is no system of ventilation. Due to this, many times losses are more. Modified containers with proper ventilation are required to decrease losses during shipment.

7.7 Inadequate Market Intelligence

Accurate information is not available at a time about price of different places internally or internationally so also the prospect of production within the country and abroad. Because of this many times it is not possible to get the stocks of particular quality at competitive price. Because of this uncertainty many times commitments made at a time are not possible to be fulfilled.

7.8 Sanitary and Phyto-sanitary Requirements for Export

The garlic grades and qualities are as follows:

a) General Characteristics

Garlic shall be obtained from varieties *Allium sativum* L. It may be fresh, semidry or dry. Garlic bulbs shall be (a) sound, fresh in appearance; (b) clean, free of any visible foreign matter; (c) firm; (d) free from damage caused by pests; (e) free from damage caused by frost or sun; (f) free from externally visible sprouts; (g) free of abnormal external moisture; (h) free of any foreign smell and/or taste. They shall comply with the residue level of heavy metals, pesticides and other food safety parameters as laid down by the Codex Alimentarius Commission for exports.

b) Criteria for Grade Designation

The grade designation requirement and grade tolerance are given in Table 3.

c) Other Requirements

The development and condition of the garlic must be such as to enable it to withstand transport and handling and arrive in satisfactory condition at the place of destination. Garlic must be presented as loose in the package, with cut stems. The length of the stem should not exceed10 cm. in the case of fresh and semi-dry garlic and 3 cm in case of dry garlic.When the garlic are presented in bunches it should indicate the number of bulbs, net weight and the stems must be evened off. In the case of dry and semi-dry produce the garlic may be presented in strings indicating number of bulbs. There should be at least six bulbs per string and mention of net weight. In the

Table 3: Grade Designation and Requirements

Grade Designation	*Grade Requirements*	*Grade Tolerances*
Extra class	Garlic shall be of superior quality. They shall be characteristic of the variety and/or commercial type. The bulbs shall be intact; regular in shape; properly cleaned. They must be free of defects, with the exception of very slight superficial blemishes, provided these do not affect the general appearance of the produce, the quality, the keeping quality and presentation the package. The cloves must be compact; The roots of dry garlic must be cut off flush with the bulb.	5 per cent by weight of Garlic not satisfying the requirements of the grade, but meeting those of Class I grade or, exceptionally, coming within the tolerances of that grade.
Class I	Garlic shall be of good quality. They shall be characteristics of the variety and/or commercial type. The bulbs shall be intact; of fairly regular shape The following slight defects, however, may be allowed provided these do not affect the general appearance of the produce, the quality, the keeping quality and presentation in the package. ☆ Slight tears in the outer skin of the bulb; cloves must be reasonably compact.	10 per cent by weight of Garlic not satisfying the requirements of the grade, but meeting those of Class II or, exceptionally, coming within the tolerances of that grade.
Class II	Garlic which do not qualify for inclusion in the higher grade, but satisfy the minimum requirements. The following defects may be allowed, provided the garlic retain its essential characteristics keeping quality and presentation. ☆ Tears in the outer skin or missing parts of the outer skin of the bulb; ☆ Healed injuries; ☆ Slight bruises; ☆ Irregular shape; ☆ Upto three cloves missing.	10 per cent by weight of Garlic not satisfying the requirements of the grade but meeting the minimum requirements.

case of presentation in bunches or strings, each package must have uniform characteristics (number of bulbs or net weight).Irrespective of the type of presentation, the stems must be cut cleanly. Package shall be marked 'fresh garlic', 'semi-dry garlic' or 'dry garlic'.

Size is determined by the maximum diameter of the equatorial section. The minimum diameter shall be at 45 mm for garlic classified in the 'Extra' Class and 30 mm for garlic classified in Classes I and II. Where the diameter of the smallest bulb is less than 40 mm for garlic presented loose with cut stems or in bunches, the difference in diameter between the smallest and the largest bulb in the same package may not exceed15 mm. Where the diameter of the smallest bulb is than 40 mm or more this differences should not exceed 20mm.Maximum permissible limit for variation for all grades of produce is 10 per cent by weight of bulbs not satisfying the requirements as regards sizing and the size indicated but conforming to the size immediately above and/or below that specified. Within these tolerances, not more than 3 per cent of bulbs may have a diameter smaller than the specified minimum but not less than 25 mm. Garlic bulb without clove (single) may have minimum diameter of 20 mm for Extra class and 15 mm for Class I and Class II.

8.0 Conclusion

Appropriate postharvest management of garlic is essential for minimizing the losses during marketing and storage. Sprouting, rooting and decay are some of the major factors responsible for huge postharvest losses. Hence, care must be taken right from harvesting at appropriate maturity, curing for removal of moisture from outer skin, packaging and storing at 0 °C with 65-75 per cent RH for long duration. However, for short term storage garlic could be stored at 27-32 °C with low RH. Similar to onion gamma irradiation of well cured garlic after harvest is useful for minimizing the storage losses due to sprouting and rooting. These treatments are also effective in checking discolouration during storage. Proper grading and use of SPS protocol are necessary for fetching better price of the produce in the export market.

9.0 Important References

Anonymous, 2012. National Horticultural Research and Development Foundation, Nasik.

Bhonde, S R. and HariPrakash 2006. Garlic cultivation in India, NHRDF, Nashik, *Tech Bul* 17, pp.50.

Bonginwar, D.R. and Shirsat, S.G.2000.Demonstration Food irradiationfacilities for prevention of losses due to sprouting in onion/garlic by use of gamma irradiation at Lasalgoan in Nasik district of Maharashtra. *National Symposium on Onion-Garlic: Production and Postharvest Management. In: Challenges and Strategies. Souvenir* pp. 33-42.

Kendler, Barry S. 1987. Garlic (*Allium sativum*) and onion (*Allium cepa*): A review of their relationship to cardiovascular disease. *Preventive Medicine* **16** (5): 670–685. doi:10.1016/0091-7435(87)90050-8. PMID 3317392.

Keys, Ancel1980. Wine, garlic, and CHD in seven countries. *The Lancet* **315** (8160): 145–146. doi:10.1016/S0140-6736(80)90620-0. PMID 6101471.

Lawande K.E. and Tripathi, P.C.2006.Marketing of irradiated onion and garlic. *Proc. of National symposium on radiation processing of foods, food products and feeds. Hyderabad,* pp.82-90.

Tripathi P.C. and Lawande, K.E. 2004. Pyajavam Lahsun Bhandaran. *Technical Bulletin No. 10* Published by NRC Onion and Garlic, Rajgurunagar pp. 20.

Tripathi, P. C. and Lawande, K. E. 2009. Effect of storage environment and packing material on storage losses in garlic. *Indian J. Hortic.* 66 (4):511-515.

Tripathi, P. C. andLawande, K.E. 2007. Effect of cold storage and gamma-irradiation on storage losses in garlic. *J. Spices and Aromatic Plants.* 17 (1):23-26

Chapter 31
Ginger

Ananta Saikia
Professor
Sanjib Sharma
Assistant Professor
Department of Horticulture, Assam Agricultural University, Jorhat – 785 013, Assam, India
E-mail: ananta.saikia@gmail.com; E-mail: sanjibsir@rediffmail.com

1.0 Introduction

Ginger (*Zingiber officinale* Rosco.) is one of the most widely used spices around the world and is used as a condiment in large number of foods and beverages. Rhizomes of ginger plants are eaten fresh or cooked as vegetables and used for flavouring foods. Ginger originated in South East Asia/Indo-Malaysian/Indo-China region. It has approximately 1400 species in 47 genera, presently grown in different geographical regions of the world.

Gingers are perennial herbs producing aromatic rhizomes that are subterranean or above ground which are rich in nutrients especially vitmins and minerals. Each rhizome can produce erect leafy shoots. Inflorescences are terminal, borne either on leafy shoots or on special erect shoots near the base of the plant. It belongs to the family *Zingiberaceae* and botanically known as *Zingiber officinale* Rosco. The family has three tribes, *viz., Zingibereae, Alpinieae* and *Hedychieae*. Zingiber is the only genus of the tribe Zingibereae. The tribe *Alpinieae* consists of the genera *Alpinia, Amomum, Elettaria, Etlingera, Elettariopsis, Geocharis, Geostachys, Hornstedtia* and *Plagiostachys*. The tribe *Hedychieae* includes *Boesenbergia, Camptandra, Curcuma, Haniffia, Hedychium, Kaempferia* and *Scaphochlamys*.

Ginger has a long and well-documented history of both culinary and medicinal use throughout the world, especially in Chinese, Indian and Japanese medicinal care. Ginger is considered as one of the world's best medicines since time immemorial which was referred to as "Viswa Bhesaj" (meaning universal medicine) and "Mahoushadha" (meaning the great medicine) in Sanskrit, during ancient Gupta era in India. It is reported to warm and soften stomach, cure pains, cough, chest disorders and other bodily discomforts. Ginger's therapeutic applications include analgesic, anti-arthritic, anti-oxidant, anti-fungal, antibacterial, anti-viral, anti-ulcer, hypo-cholesteremic, cardiotonic, stimulant, aphrodisiac and anti-helminthic, without adverse side-effects.

2.0 Status of Cultivation

The major ginger growing countries are India, China, Nepal, Nigeria, Indonesia, Jamaica, Taiwan, Japan, Sierra Leone, Cameron, Bangladesh, Philippines, Sri Lanka, Thailand, Malaysia, Costa Rica, Côte d'Ivoire, Ethiopia, Uganda, Tanzania, Trinidad and Tobago, Fiji, Mauritius, Republic of Korea and Dominican Republic. India is a leading producer of ginger in the world and, during 2012, the country produced 703 thousand tonnes of the rhizomes from an area of 150 thousand hectares. During the year, India contributed 27 per cent to the global production of 2095 thousand tonnes from 322.16 thousand hectares of cultivated land distributed in various countries. During last three decades, India continues to be the leading ginger producing country in the world (Figure 1).

Ginger is cultivated in most of the states in India. According to National Horticulture Board (NHB) the area under ginger in India is 136.25 thousand ha with a production of 682.63 thousand tonnes with productivity of 5.01 t/ha during 2012-13. Assam is the leading the state with production of 136.83 thousand tonnes from 17.84 thousand ha and productivity of 7.67 t/ha. Assam, Gujarat, Karnataka, Arunachal Pradesh, Meghalaya, Sikkim, Kerala and Mizoram together contributed more than 70 per cent towards the country's total production during 2012-13. Gujarat is leading among all the states with a productivity of 16.06 t/ha (Table 1).

Production area of ginger is mainly dominated by the North Eastern and southern states in India. In NER, ginger is generally cultivated in the *jhum* fields in all the hilly states except Sikkim and mainly as a rain-fed crop. Ginger is grown both as rainfed and irrigated crop in other parts of the country. The details of concentrated pockets of ginger in different states are given below (Table 2).

3.0 Major Commercial Varieties

3.1 Local Cultivars and Land Races

Cultivars are usually named in local languages in a geographical region where they are commonly grown and adapted by the farming community. Data presented in Tables 3–5 indicate certain important quality parameters with respect to yield, dry matter, crude fibre, oleoresin and essential oil content of both indigenous and exotic cultivars. The indigenous cultivars are Ernad, Wynad, Kuruppampadi, Thodupuzha, Narasapattam, Bhola, Jati, Maran, Thingpui, Thinglaidum, Tura, Himachal, Karakkal and Nadia. Few exotic varieties such as Rio-de-Janeiro from Brazil, Taffingiva from

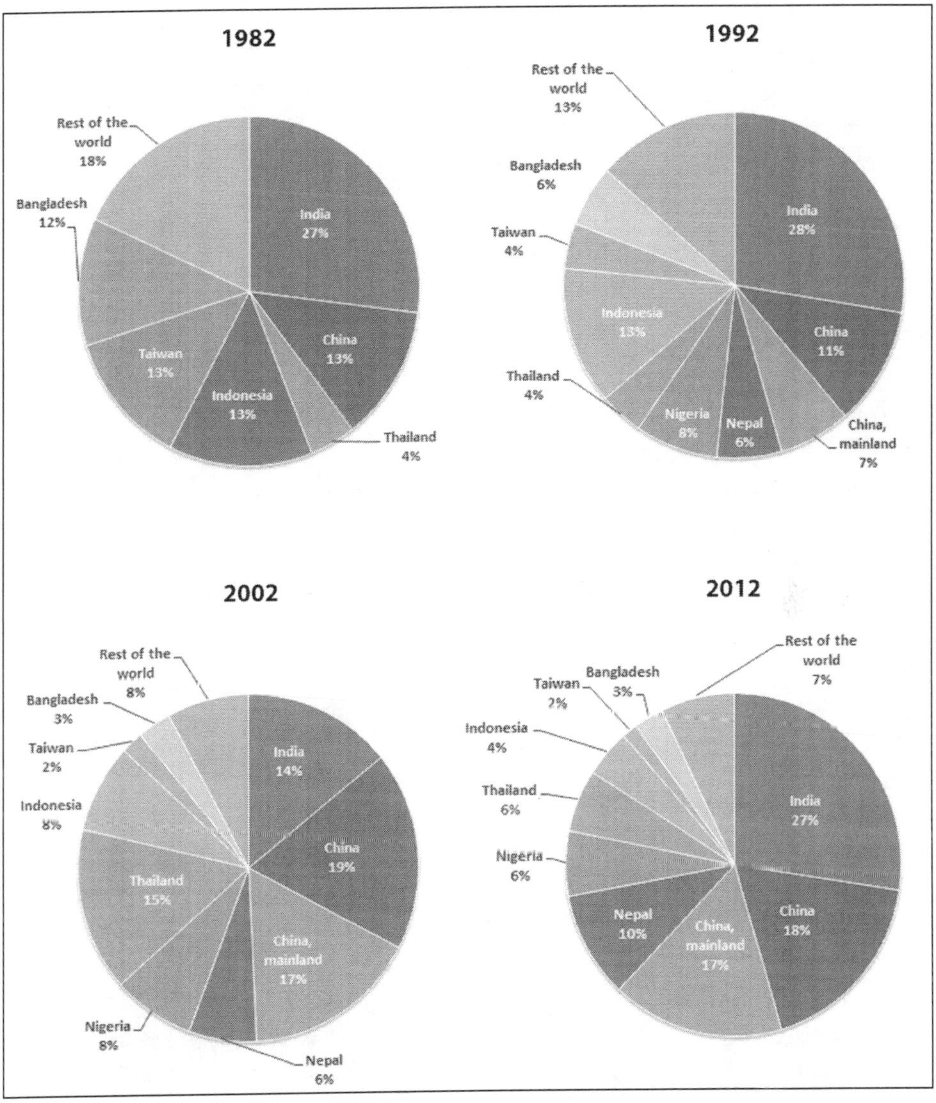

Figure 1: World Productivity of Ginger between 1982 to 2012 (FAO)

Nigeria and China from Peoples Republic of China had been introduced in India which later became popular in different parts of the country.

3.2 Improved Varieties

Distinct variations in plant morphology, rhizome characteristics and quality are also observed among the ginger cultivars grown in India. Continuous research efforts by the different organization have led to the development of many improved

Table 1: Area, Production and Productivity of Ginger in India

States	*2012-13*		
	Area ('000 ha)	*Production ('000 t)*	*Productivity (t/ha)*
Andaman and Nicobar	0.22	1.91	8.68
Andhra Pradesh	2.32	20.00	8.62
Arunachal Pradesh	7.00	57.00	8.14
Assam	17.84	136.83	7.67
Bihar	0.56	0.84	1.5
Chhattisgarh	1.91	2.32	1.21
Gujarat	4.40	70.65	16.06
Haryana	0.45	5.65	12.56
Himachal Pradesh	2.10	3.74	1.78
Jammu and Kashmir	0.03	0.04	1.33
Karnataka	16.51	57.80	3.50
Kerala	6.36	33.75	5.31
Madhya Pradesh	9.00	15.00	1.67
Maharashtra	1.06	1.04	0.98
Manipur	2.40	3.84	1.60
Meghalaya	9.44	56.80	6.02
Mizoram	7.28	28.39	3.90
Nagaland	5.32	36.00	6.77
Odisha	15.84	35.00	2.21
Rajasthan	0.15	0.31	2.07
Sikkim	9.25	51.56	5.57
Tamil Nadu	0.33	4.10	12.42
Tripura	1.80	7.60	4.22
Uttar Pradesh	0.82	4.02	4.90
Uttarakhand	2.36	23.44	9.93
West Bengal	11.50	25.00	2.17

Source: Indian Horticulture Database, NHB 2013.

varieties suitable for growing in different agro-climatic zones of the country. Some of the prominent improved varieties are Suruchi, Surabhi, Suprabha, Himgiri, Manantoddy, Cochin ginger (NUGC), Calicut ginger (NUBK), IISR Varad, IISR Mahima and IISR Rajitha. Table 4 and 5 are show some popular ginger varieties with specific attributes are presented in Table 6.

Table 2: Concentrated Pockets of Ginger in India

State	*Districts*
Arunachal Pradesh	Dibang valley, East Siang, Lohit, Tirap, Upper Siang
Assam	Borpeta, Darrang, Dibrugarh, Golaghat, Kamrup, Karbi Anglong, N.C. Hills, Sivasagar, Sonitpur, Tinsukia
Kerala	Alappuzha, Idukki, Palakkadu, Wayand
Meghalaya	East Garo Hills, Ri-Bhoi, West Garo Hills
Odisha	Kondhamal, Koraput
Sikkim	East Sikkim, South Sikkim

Table 3: Physico-chemical Characteristics of Local Cultivars/Land Races of Ginger

Cultivar	*Fresh Mean Yield (t/ha)*	*Maturity (Days)*	*Dry Recovery (per cent)*	*Crude Fibre (per cent)*	*Oleoresin (per cent)*	*Essential Oil (per cent)*
Assam	11.78	210	18.0	5.8	7.9	2.2
China	9.50	200	21.0	3.4	7.0	1.9
Himachal	7.27	200	22.1	3.8	5.3	0.5
Maran	25.21	200	20.0	6.1	10.0	1.9
Nadia	28.55	200	22.6	3.9	5.4	1.4
Rio-de-Janerio	17.65	190	20.0	5.6	10.5	2.3

Source: Sasikumar *et al.*, 2009.

Table 4: Physico-chemical Characteristics of some Improved Varieties of Ginger

Cultivar	*Mean Yield t/ha*	*Dry Recovery (%)*	*Oil Content (%)*	*Oleoresin (%)*	*Crude Fibre (%)*	*Mean Days to Maturity*
Himgiri Clonal selection from Himachal Local	14.0	20.6	1.6	4.3	6.0	230
IISR Mahima Selection from germplasm	23.2	23.0	1.7	4.5	3.3	200
IISR Rajitha Selection from germplasm	22.4	19.0	2.4	6.2	4.0	200
IISR Varada Selection from germplasm	22.6	19.5	1.7	6.7	3.3	200
Suprabha Selection from Kunduli local	16.6	20.5	1.9	8.9	4.4	230

Contd...

Table 4–*Contd...*

Cultivar	*Mean Yield t/ha*	*Dry Recovery (%)*	*Oil Content (%)*	*Oleoresin (%)*	*Crude Fibre (%)*	*Mean Days to Maturity*
Suravi Selection from germplasm	17.5	23.0	2.1	10.2	4.0	225
Suruchi Induced mutant of Rudrapur local	11.6	23.5	2.0	10.0	3.8	220

Source: Ravindran and Babu, 2005.

Table 5: Varietal Characteristics of some Commercially Grown Ginger Varieties

Cultivar	*Characteristics*				
	Rhizome Size	*Skin Colour*	*Pungency*	*Flavour*	*Fibre*
Maran	Medium	Buff	Pungent	High	Less
Nadia	Medium to Bold	Yellowish	Moderately pungent	Mild	Less
Rio-de-Janeiro	Bold	Buff	Pungent	High	Less
Thingpui	Medium	Buff	Pungent	High	Less
Wynad	Bold	Buff	Pungent	Moderate	Less

Table 6: Ginger Cultivars with Specific Attributes

Attributes	*Cultivars*
High yield	Mahima, Rajatha, Suprabha, Suravi and Varda
High dry recovery	Jhorat local, Kuruppampadi, Mahima, Maran and Varda
High oil and oleoresin contents	Ernad, Kunnamangalam and Wynad
High gingerol and shogaol	Ambalavayal, Kunnamangalam, Rio de Janeiro, Sweetning Puri, and Wynad

Kerala is leading in exporting dry ginger 'Cochin ginger' and 'Calicut ginger' or 'Wynad ginger'. Malabar (Kerala) ginger is rich in volatile oil. Assamese ginger and Monantody ginger are rich in oleroresins. The local strains from Pouni and Choki are relatively better in terms of yield effectively resisting disease problems. The strains from Tanda Ganderwan are suitable for preparing '*sonth*' rhizomes.

4.0 Nature and Causes of Postharvest Losses

4.1 Postharvest Diseases

Loss from rot caused by pathogens begins in the field and continues in the packing house, cold storage and refrigerated containers on ships, causing significant losses for the ginger industry. Postharvest disease in ginger is normally due to improper handling during postharvest period which result in injury to the rhizome. Keeping ginger in storage environment a favourable to pathogens, accelerates the postharvest decay. Careful harvesting and handling practices, sanitation of the cleaning water, curing of the rhizomes, application of a postharvest fungicide etc. can reduce postharvest decay of ginger. During storage of ginger, the dominant pathogenic fungi are *Fusarium oxysporum, Pythium deliense, P. myriotylum,* and other (saprophytic) fungi such as *Geotrichum candidum, Aspergillus flavus, Cladosporium lennissimum, Gliocladium roseum, Graphium album, Mucor racemosus, Stachybotrys sansevieriae, Thanatephorus cucumeris,* and *Verticillium chlamydosporium.*

a) Bacterial Soft Rot

Bacterial soft rot in ginger is caused by *Erwinia carotovora* and is considered the principal bacterial storage disease of ginger. These bacteria enter the ginger rhizome through wounds in the tissue. Symptoms include a soft wet rot of the tissue, which has a strong foul odour. The disease spreads rapidly under warm and humid conditions.

b) Blue Mold

Blue mould, caused by the fungus *Penicillium*, is a very common postharvest rot of ginger when the rhizome are stored at a high RH or in less ventilated store or when come into contact with another moldy product. Moisture condensation on the rhizome surface encourages blue mould. This is a mild surface infection and is superficial only and does not lead to decay. But it imparts an unacceptable appearance to the rhizomes. However, if the mold develops on cut ends and injured areas, it will typically result in internal decay resulting in pale to dark brown, and firm or soft tissue. The decay is accelerated at temperature between 15 °C and 20 °C.

c) Dry Rot

Dry rot, caused by the soil-borne fungus *Fusarium oxysporum* Schl., is another common postharvest disease of ginger. The affected rhizomes do not show any external symptoms. Browning of internal tissues and yellow colouration occur in the pith region of the rhizome. The buds on the rhizomes become black. This rot can cause serious problems invading rest of the rhizome that becomes brown, dry and shrivelled. Its infection appears when rhizomes are wounded and damaged by insects or nematodes. External symptoms of dry rot include off-coloured dry sunken lesions, typically bordered by a brown margin, on the rhizome surface. Under high humid condition, the ginger surface is covered with a dense white mould.

d) Red Rot

The red rot of the rhizome is caused by *Verticillium chlamydosporium* Goddard and shows superficial mycelial growth with the internal tissues turning red.

e) *Sclerotium* Rot

Sclerotium rot, caused by the soil-borne fungus *Sclerotium rolfsii* Sacc., commonly appears on rhizomes harvested from stressed plants. The decay starts with a well-defined separation between rotten and healthy portions of the rhizome. Gradually, a white mold covers the ginger surface and the rhizomes start shrinking. Small spherical white fungal resting bodies, about 1 to 2 mm in diameter, develop on the mould which later turn brown.

f) Soft Rot

The soft rot or *Pythium* rot, caused by *Pythium ultimum* Trow, *P. aphanidermatum* (Edson) Fitzp., *P. vexans* de Bary, *P. deliense* Meurs and *P. myriotylum* Drech., has also been reported in which rhizomes become brown to black in colour, soft and watery. This is a soil-borne fungus that can be severe on ginger harvested during the rainy season or on rhizomes grown in poorly drained soils. The buds become brown in colour with mycelial growth on its surface. As the decay progresses, the tissue breaks down into a soft and watery mass. Fungicides, at this stage, are not recommended, but the disease can be reasonably controlled if the rhizomes are adequately cured and stored at 12-14 °C.

g) Watery Rot

Watery rot, caused by the fungus *Rhizopus*, is one of the most rapidly developing storage rots of ginger. A water-soaked lesion starts to develop a few hours after the root is wounded or bruised during harvest or handling. In 4 to 5 days, the entire inner tissues of the root become mottled, soft, slimy and watery as they liquefy and the infected portion is soon covered with large amounts of white mould which will eventually turn black. *Rhizopus* is a wound pathogen. Soft rot caused by *Rhizopus* has some similarities with bacterial soft rot, as both produce the watery, flesh decaying rot and both attract fruit flies. However, bacterial rot produces an unpleasant smell while *Rhizopus* rot emits a pleasant fermented odour that attracts fruit flies. Also, bacterial rot does not produce whiskers unlike *Rhizopus* soft rot.

h) Postharvest Treatments

Chemical treatments before storage of ginger rhizomes can prevent many of the above rots. Postharvest (prestorage) application of benomyl (750 ppm *a.i.*) and/or gibberellic acid (150 ppm) can be applied to control the rots. A combination of mancozeb and carbendazim treatment to rhizomes controls storage rot of ginger. *Trichoderma viridi* and *T. harzianum* have also been effectively used to control storage diseases of ginger.

4.2 Postharvest Disorders

a) Chilling Injury

Ginger rhizomes should not be stored at a low temperature below 12 °C, as it may cause chilling injury. This injury is manifested in the form of pitting and sunken lesions on the rhizome surface, shrivelling, softening, flesh darkening and decay. Those shorten the shelf life and lower down the eating quality of ginger.

b) Sprouting

During storage, ginger rhizomes start sprouting if the temperature goes above 15.6 °C and the rate increases with increase in temperature. Keeping the rhizomes at

12-14 °C will prevent sprouting. No effective chemical has been reported as sprout inhibitor for ginger. Use of gamma radiation has been so far effective in controlling sprouting. Dipping washed and dried rhizomes in *Trichoderma* suspension (10^8 spores ml^{-1}) followed by air-drying, and packing in 250 gauge LDPE bags, and finally exposing to 50-60 Gy gamma radiation can inhibit spouting in the rhizomes.

4.3 Preharvest Crop Management

Various preharvest factors including management practices and environmental factors influence postharvest quality. The quality of the product during the postharvest phase can be ensured if proper preharvest management practices are followed. It is possible to get high market prices, create employment opportunities and sustainable supply only through the use of appropriate variety, improved cultural practices, irrigation, pest and disease management etc.

4.3.1 Cultural Practices

Selection of variety is crucial which determines yield and quality of ginger crop. Thus, one should study all the available varieties critically, based on the anticipated end use of the produce, before selecting a variety for cultivation. It is, thus, essential to encourage producers to use appropriate varieties to get economically rewarding crop, increase income level and make more competition in the markets. The best time for planting ginger in the West Coast of India is during the first fortnight of May as soon as the premonsoon showers start. The usual time to start the crop in the North Eastern states is February-April. Under irrigated conditions, it can be planted well in advance during the middle of January or early February.

There is an increasing demand for organically-grown ginger both in fresh form as well as dried products. A few useful tips for organic ginger are as enumerated below.

i) A buffer zone of 20-25 feet all around from conventional farm should be maintained and a two year conversion period should be adopted.

ii) Tillage operations during land preparation should be kept to minimum.

iii) Soil should be solarised before planting of ginger rhizomes.

iv) Pest and disease free planting material should always be sourced from organically grown ginger belts which must be treated with Trichoderma (@ 2g/kg seed) both before storage and planting.

v) Application of FYM or quality compost (5-10 tonnes), groundnut cake (350 kg), neem cake (1700 kg), rock phosphate (250 kg), bio fertilizer (20 kg) and 1 tonne wood ash in one hectare area at the time of planting should be practised.

vi) Mulching may be done with green leaves @10 tonnes/ha at time of planting and @ 5 tonnes/ha each after 40 and 70 days of planting. Cow dung slurry or liquid manure could be poured on beds after mulching.

vii) Neem oil (0.5 per cent) should be sprayed at two week intervals to control insects. The caterpillars of shoot borer can be hand picked and destroyed.

4.3.2 Removal of Mother Rhizome

Traditionally, farmers of North East India in both plains and hills use bigger size rhizomes at the time of planting. They find it economical although this practice initially increases the seed rate to 2.0 – 2.5 t/ha against the normal rate of 1.5t/ha. Most of the farmers of the region remove the mother rhizome (Plate 2) when ginger crop attains 60 days age or 3-4 leaves stage leaving the sprouted piece of rhizome in the soil. This practice provides the farmer an additional yield of about 10 q/ha which is sold during off-season at much higher rates. After extraction of the mother rhizomes, FYM is applied and plants are earthed up. It also believed that removal of the mother rhizome gives sufficient space to the developing daughter rhizomes for growth.

5.0 Postharvest Management

5.1 Maturity Indices

Traditionally, farmers harvest gingers as soon as the foliage senesces and rhizomes attain considerable size. Gingers are harvested once the foliage begins to turn yellow and starts drying. This also indicates maturity of rhizomes. This natural process of aging, accompanied by foliage yellowing, typically sets in 7 to 9 months after planting. Delaying harvest till all of the leaves are dried up is not recommended as it reduces rhizome quality, storage life, increases fibre content and incidence of sprouting.

It is best to harvest rhizomes of ginger based on the end use. The rhizomes should be harvested, while they are still tender, medium in pungency and low in fibre content before they are fully mature for consuming as fresh or for making preserves. Thus, green ginger can be harvested right from five months onward. For making dried spices and extracting oils and oleoresins, it is better to harvest ginger at full maturity when the leaves turn yellow; however, harvesting beyond that stage may lead to reduced pungency and oil content with subsequent increase in fibre content in the rhizomes.

As per FAO's guidelines, the best harvest time for fresh consumption is 5 months after planting. For production of ginger preserve and making dried ginger or extraction of ginger oleoresins or essential oils, the rhizomes should be harvested between 5-7 months and 8- 9 months after planting when the leaves start yellowing.

When market demand for ginger is high with rise in price, farmers occasionally harvest gingers at the premature stage as it seem it to be cost effective. In this situation, one should sacrifice some randomly selected plants and dug up to assess the average rhizome size. However, it should be kept in mind that the early harvest of non-senescent plants can result in more weight loss and decay. The skin of early harvested ginger is very tender and the fresh wound at the point of separation between the stem and rhizome will take more time to heal.

5.2 Harvesting

Harvesting of ginger in all the ginger producing states of India is carried out during November to January. The rhizomes are harvested by lifting the clumps carefully with a spade or digging fork or hoe by loosening the soil around the crown of the plant. Mechanical digging devices are also available for use on large plantations. Initially, a significant portion of the senescent foliage should be removed to make the rhizomes more accessible. The process should be done carefully in order to avoid damaging the rhizomes which have delicate skins and are easily wounded during harvest and handling. The rhizomes are then carefully separated from the plant. The rhizomes in a clump are highly irregular in shape and close to each other which favours the inclusion of soil and dirt in the interspaces. These should be removed by thorough washing. If needed, the interspaces are pricked clean using pointed bamboo stick. The adhering roots are also to be trimmed. After washing, the ginger is lightly dried in the sun or under a hot air blower before being marketed in this form or to be used for further processing. Young rhizomes harvested earlier will still have an actively growing stem which should be cut off slightly above the point of attachment to the rhizome. Harvesting during very wet or very dry conditions is not advisable as this will increase the amount of skinning and make digging of rhizomes from the soil considerably difficult.

While field cleaning, one should wear cotton gloves for hygiene and to prevent infection. The rhizomes should never be thrown or dropped during postharvest operations. The rhizomes should be graded in the field itself and any unmarketable, damaged, or diseased rhizome should be discarded. The rhizomes should never be exposed to direct sunlight for more than 30 minutes to prevent desiccation. The harvested, cleaned and graded gingers should be placed directly into strong well-ventilated wooden or plastic field crates for transfer from the field. Transporting ginger in mesh sacks will increase the level of skin damage due to rubbing of the delicate skin against the mesh.

5.3 Pretreatments

Gingers are marketed as fresh rhizomes, whole dry form, bleached rhizomes, powder, flakes, paste, oils and oleoresins and other processed forms like preserves, pickles etc. Considering the above standards, one should adopt proper postharvest operations to improve the quality parameters of fresh and dry gingers to meet the above requirements.

5.3.1 Precooling

Generally, ginger should be harvested in the morning or the afternoon so that the operation is performed at the coolest possible temperature. The ginger rhizomes, immediately after harvest, should be precooled to remove the field heat. Preferably, forced-air or room-cooling to 12 to 14 °C should be used. In forced air cooling, refrigerated air is forced into the harvested rhizomes kept in boxes or crates. Several airflow systems are available, but tunnel cooler is common. Room cooling of the rhizomes can be done by keeping them in a refrigerated room. Alternately, if these facilities are not available, ginger rhizomes can be precooled by keeping them spread over clean concrete floor of an airy room.

5.3.2 Cleaning

Relatively clean rhizomes for sale in the domestic market may not require any further cleaning. However, thorough and proper cleaning is required for gingers intended for export. The rhizomes should also be properly cleaned for long-term storage immediately after harvest and then cured. Ginger should be scrubbed by hand or with a soft-bristled brush in clean water sanitized with 150 ppm hypochlorous acid. The rhizomes are to be washed as soon as possible after harvesting, as the adhering soil is more difficult to remove once dried (Plate 3). Pressure washing using water jet is commonly advised as it efficiently cleans the rhizomes subsequently reducing the microbial load.

One should be careful enough to prevent rhizome breakage during handling. Broken, split, or bruised rhizomes can easily shrink due to moisture loss and are prone to decay. The pH of the cleaning water is to be maintained at 6.5 for maximising the sanitizing effect of the hypochlorous acid. The water needs to be changed after it builds up with soil particles and the strength of the hypochlorous acid goes down 150 ppm. A fungicide treatment, preferably with benomyl (500 ppm *a.i.*) or thiabendazole (1000 ppm), is also recommended and can be applied in the wash tank or as a separate overhead spray application after cleaning. After the cleaning process, the ginger should be air dried and sorted.

Specific equipment is suggested for optimum cleaning of the dried rhizomes. An air screen separator will help to remove dead insects, excreta and extraneous matter, while a rotary knife cutter with a screen separator will remove residual insects and other extraneous matter. The cleanliness specifications set by American Spice Traders Association are as followed in which the level of contaminants permitted must fall below the values shown (except for 'Whole insects, Dead' which cannot exceed the limit shown) in Table 7:

Table 7: Cleanliness Specifications for Ginger

Parameters	*Value*
Whole insects, Dead (by count)	4.0
Excreta, Mammalian (by mg/lb)	3.0
Excreta, Other (by mg/lb)	3.0
Mold (per cent by weight) infested pieces by weight	More than 3 per cent moldy pieces and/or insect
Insect defiled/infested (per cent by weight)	More than 3 per cent moldy pieces and/or insect infested pieces by weight
Extraneous/foreign matter (per cent by weight)	1.0

Source: APEDA 2014

The fresh gingers shall have to comply with the maximum levels of the Codex General Standard for Contaminants and Toxins in Food and Feed (CODEX STAN 193-1995) and the maximum residue limits for pesticides established by the Codex Alimentarius Commission. It is recommended that the produce covered by the provisions of this Standard be prepared and handled in accordance with the appropriate sections of the Recommended International Code of Practice – General Principles of Food Hygiene (CAC/RCP 1-1969), Code of Hygienic Practice for Fresh

Fruits and Vegetables (CAC/RCP 53-2003), and other relevant Codex texts such as Codes of Hygienic Practice and Codes of Practice.

The produce should comply with any microbiological criteria established in accordance with the Principles for the Establishment and Application of Microbiological Criteria for Foods (CAC/GL 21-1997). Quality specifications are imposed by the importing country, and pertain to cleanliness specifications. Hence, suitable postharvest processing care must be taken to meet minimum requirements to avoid rejection. The rhizomes should be disinfected with ethylene oxide or irradiation before export. Before doing so, the exporter must be well-informed about the standards set by the importing country.

5.3.3 Curing

Ginger rhizomes, after cleaning, should be cured by air drying at ambient temperature (22 °C to 26 °C) and 70-75 per cent RH for several days to allow the skin to thicken and the cut surfaces to suberize. Curing helps in reducing postharvest weight loss and decay. After curing treatment, the ginger should be put in well-ventilated containers for long-term storage.

5.4 Sorting

After discarding damaged and injured rhizomes, the marketable portion should be sorted according to size and consumer acceptability. The fresh rhizomes for marketing should be clean, bright yellow brown in colour, and appear fresh without any symptom of desiccation or sprouting, bacterial or fungal infection and objectionable skin blemishes.

As per the Codex Standards (CODEX STAN 218-1999) outlined by the Codex Alimentarius Commission, fresh whole ginger must be: "whole, firm, sound, produce affected by rotting or deterioration such as to make it unfit for consumption is excluded, clean, practically free of any visible foreign matter, practically free of damage caused by pests affecting the general appearance of the produce, free from abnormal external moisture, and properly dried if washed excluding condensation following removal from cold storage, free of any foreign smell and/or taste, free from abrasions provided light abrasions which have been dried properly are not regarded as a defect, sufficiently dry for the intended use; skin and stems and cuts due to harvesting must be fully dried".

Additionally, the development and condition of the ginger must be such as to enable it to withstand transport and handling and to arrive in satisfactory condition at the place of destination.

There are no minimum or maximum size limits for domestic marketing of gingers. However, the minimum rhizome size for export is 150 g with no maximum size limit. The main stem of the rhizome should not be less than 3 cm thick and 12 cm long. Rhizomes should be large, well-formed with less than 4 side branches. The internal flesh should be firm and uniformly coloured, without any indication of darkening.

5.5 Grading

As per Codex Standards, following classes of fresh gingers are defined as:

a) Extra Class

Ginger in this class must be of superior quality. It must be characteristic of the variety and/or commercial type. The roots must be cleaned, well-shaped and free of defects, with the exception of very slight superficial defects, provided these do not affect the general appearance of the produce, the quality, the keeping quality and presentation in the package. Quality tolerance limits to five percent by number or weight of ginger not satisfying the requirements of the class, but meeting those of Class I or, exceptionally, coming within the tolerances of that class.

b) Class I

Ginger in this class must be of superior quality. It must be characteristic of the variety and/or commercial type. The roots must be cleaned, well-shaped and free of defects, with the exception of very slight superficial defects, provided these do not affect the general appearance of the produce, the quality, the keeping quality and presentation in the package. Slight skin defects due to rubbing provided they are healed and dry, and the total surface area affected does not exceed 10 per cent. Quality tolerance limits to ten percent by number or weight of ginger not satisfying the requirements of the class, but meeting those of Class II or, exceptionally, coming within the tolerances of that class.

c) Class II

This class includes ginger which does not qualify for inclusion in the higher classes, but satisfy the minimum requirements specified above. The roots should be reasonably firm. The following defects, however, may be allowed, provided the ginger retains their essential characteristics as regards to quality, the keeping quality and presentation. Skin defects due to rubbing are accepted in this class provided they are healed and dry and the total surface area affected does not exceed 15 per cent. Early signs of sprouting (not more than 10 per cent by weight by unit of presentation), slight markings caused by pests, healed suberized dry cracks, slight traces of soil are permitted in this class. Quality tolerance within this class limits to ten percent by number or weight of ginger satisfying neither the requirements of the class nor the minimum requirements, with the exception of produce affected by rotting or any other deterioration rendering it unfit for consumption.

Codex Standards have also classified fresh ginger rhizomes into three grades based on size, *viz.*, A – 300g in weight; B – 200g in weight and C – 150g in weight.

According to Indian Standards (BIS, 14), there are six grades for dry gingers *viz.*, *i*) Garbled, non-bleached Calicut (NGK), *ii*) Ungarbled, non-bleached Calicut (NUGK), *iii*) Garbled, non-bleached Cochin (NGC), *iv*) Ungarbled, non-bleached Cochin (NUGC), *v*) Garbled, bleached Cochin (BGC) and *vi*) Ungarbled, bleached Cochin (BUGC).

5.6 Packaging

According to Codex Standards, "Ginger must be packed in such a way as to protect the produce properly. The materials used inside the package must be new, clean, and of a quality such as to avoid causing any external or internal damage to the produce. The use of materials, particularly of paper or stamps bearing trade specifications is allowed, provided the printing or labelling has been done with non-toxic ink or glue. Ginger shall be packed in each container in compliance with the Recommended International Code of Practice for Packaging and Transport of Fresh Fruits and Vegetables (CAC/RCP 44-1995)" (Anonymous, 3).

Bulk fresh rhizomes can be packed in jute sacks, plastic crates, wooden boxes or lined corrugated cardboard boxes for domestic sale. According to Indian Standards (BIS, 14), the whole dried ginger should be packed in "sound double jute bags or in single jute bags with waterproof lining". The bags may be lined inside with cloth or with LDPE. The ground ginger should be packed in clean and sound tin plate or glass containers or paper cartons properly lined with waterproof paper or in moisture-proof bags made of a material which does not impart any smell to the ginger ground.

5.7 Storage

5.7.1 Fresh Consumption

Healthy undamaged fresh ginger rhizomes can be stored for several months in a cold and humid environment if the correct postharvest handling and storage procedures are followed. The ideal environment for storing mature ginger rhizomes is 12 to 14 °C with 85 to 90 per cent RH in a cold storage. At this condition, the rhizomes remain in marketable condition for at least 3 months, depending on initial rhizome vigour and condition. Condensation over the rhizomes should not be allowed as it can encourage superficial mold growth. Lower RH in the storage environment leads to extensive dehydration and a wilted appearance.

Cold storage may not always be available in the production areas. In that case, a "zero energy cool chamber", designed by IARI, New Delhi, can be used to preserve fresh ginger, which can lower down the storage temperature by 6 to 7 °C than the outside temperature. In the North Eastern states, the model developed by the Assam Agricultural University, a modified version of IARI's "Zero energy cool chamber", is more suitable for ginger storage.

Use of gamma irradiation at 50-60 Gy has been proved beneficial in inhibiting sprouts and extending shelf life of fresh ginger, although loss of volatile compounds have been reported ater 5 months of storage. A combination of bio-control with *Trichoderma* sp. and storage in polyethylene bags at 25-30 °C has been reported to control the fungal rot during storage caused by *Sclerotium rolfsii* and prevented weight loss from dehydration.

5.7.2 Seed Purpose

Storage of seed rhizomes is also equally important for good germination in the next season. According to Spices Board of India, bold and healthy rhizomes from disease free plants are to be selected immediately after harvest for storing as seeds.

For this purpose, healthy and disease-free clumps are marked in the field when the crop is 6-8 months old and still green. Those selected plants are harvested upon maturity and the seed rhizomes are treated with a solution containing quinalphos 0.075 per cent and mancozeb 0.3 per cent for 30 minutes and dried under shade. After treatment, the seed rhizomes are stored in pits of convenient size under shade. The walls of the pits should be plastered with cow dung paste. The seed rhizomes are placed in pits in layers covered with a 2 cm thick layer of dry sand/saw dust. In this manner, rhizomes are placed in alternative layers in the pit leaving sufficient space at the top for adequate aeration. The pits are then covered with wooden planks keeping one or two small openings for aeration. The seed rhizomes in the pits should be inspected once in three weeks to remove shrivelled and diseased rhizomes. The seed rhizomes can also be stored in pits dug in the ground under shade. Spices Board of India also recommends incorporation of dried leaves of nux-vomica (*Strychnos nux-vomica*) into the rhizome and sand/saw dust layers to prevent infestation of rhizome scale.

5.8 Transportation

The fresh rhizomes should be transported promptly in clean and dry conditions. When available, refrigerated trailers and van containers should be used to transport gingers if the transit period is more than a week. The gingers are to be placed in clean baskets, dry sacks, trailers, hoppers or other well-aerated containers and carried to a central point for transport to the processing facility. All containers used at harvest should be cleaned regularly, especially before transporting a lot, to avoid contamination from previously-harvested plant products and other foreign matter. The containers should be kept in dry conditions, when not in use, in an area that is protected from insects, rodents, birds and other pests, and inaccessible to livestock and domestic animals. Vehicles used for transporting bulk plant materials from the place of production to storage for processing should be cleaned between loads.

6.0 Domestic Marketing and Export

Generally producers manually transport ginger and deliver to the collectors. The collectors are responsible to transport gingers to the wholesalers and the wholesalers, in turn, are transporting ginger to the exporters. The exporters are responsible to transport ginger up to the Indian port of loading. Improper handling during loading and unloading of the packages at high temperature from the bus, truck and van adversely affects the quality of ginger. The bumpy roads during transportation aid to the loss of quality of ginger. The Codex Alimentarius Commission has recommended code of practice for packaging and transport for fresh fruits and vegetables (CAC/RCP 44-1995) which applies to ginger also.

India is presently leading the ginger production in the world. More than 50 per cent of total ginger production takes place in North East India, Uttarakhand and West Bengal states. In many of these states, ginger is cultivated traditionally in organic environments. Several traditional varieties/cultivars have been used in cultivation in different states which are suitable for producing various products of ginger. In addition to these, Indian Institute of Spices Research, Indian Council of Agricultural Research, Indian Institute of Horticultural Research and many state agricultural universities are continuously improving the existing varieties along with developing new, improved and high yielding varieties. Agricultural and Processed Food Products Exports Development Authority (APEDA) has earmarked several Agri Export Zones for fresh ginger in Assam, Orissa and Sikkim. Spices Board of India provides further support for expansion of area under ginger. All the above facts contribute to strengthen India's position as a strong player in ginger export (Table 8).

The following measures can increase India's competitiveness in ginger exports:

i) India's main competitors in ginger export are China, Indonesia, Nepal and Thailand. India needs to percolate its superior and high yielding cultivars like IISR Varada etc. in states like Assam, Mizoram, Meghalaya, Sikkim and Orissa, to bring sizeable area under superior cultivars.

ii) With increased productivity, the cost of production will go down. This will help India to provide better quality ginger at competitive prices.

iii) In international market, there is a demand for clean and bigger finger sized ginger. The gingers produced in North Eastern states are plump with less fibre and medium pungency having clump size above the Codex size standard for fresh ginger. Thus, India must establish postharvest handling facilities for cleaning, grading, postharvest treatment and appropriate packing of ginger for export.

7.0 Conclusion

Although, India is the leading ginger producing country in the world, the productivity is low as compared to competing countries like China. Many constraints are limiting the production and productivity of ginger, chief among them being the farmers' adherence to traditional varieties in spite of availability of high yielding varieties, pests and diseases, and improper postharvest management. The rhizome of ginger plant is used both as spice and medicine. India occupies a predominant position in ginger production, contributing 36.5 per cent of the total world production. This is due to the fact that ginger is grown in almost all the states of India owing to versatility in adoption for climate and soil. Harvesting time of ginger varies depending on location, but largely, ginger is harvested between December to March. However, the demand of the fresh ginger remains constant by the consumer throughout the year. The self-life of ginger is limited and moisture loss, sprouting, rotting, rooting which results in weight loss and shriveling. Packaging of ginger plays a very crucial role in postharvest handling and storage. In India the normal practice is to pack the ginger rhizomes in 55kg gunny bags during transportation. Hence, use of suitable packaging including MAP may have a great challenge in reduction of huge postharvest losses which sometimes goes up to 50 per cent due to lack of awareness, distance between production and market place, inaccessible road condition, poor transport vehicles etc.

Table 8: Country-wise Export of Ginger from India

Qty. in MT, Value in Lakhs

Countries	*2009–10*		*2010–11*		*2011–12*		*Per cent Share of Value*
	Qty.	*Value*	*Qty.*	*Value*	*Qty.*	*Value*	
Bangladesh	6276.74	1170.06	12647.51	2997.16	13062.24	2383.54	8.95
Egypt	50.00	48.43	218.02	311.54	728.02	843.88	3.17
Germany	146.03	161.37	297.21	437.14	1375.37	1964.74	7.38
Morocco	280.00	388.53	929.16	1569.42	1361.33	1684.86	6.33
Nepal	233.01	137.40	77.93	60.51	869.11	1083.47	4.07
Saudi Arab	727.09	767.75	997.56	1497.91	1675.60	2028.53	7.62
Spain	348.35	479.45	273.17	481.10	808.08	1032.44	3.88
UAE	414.50	301.32	741.42	790.48	2249.78	2524.77	9.48
United Kingdom	642.40	639.21	646.75	845.23	733.64	1120.72	4.21
USA	904.52	718.38	2302.42	2093.49	4303.35	3963.02	14.88
Other Countries	2152.27	2096.20	5498.94	5162.59	8449.88	7996.78	30.03
Total	12174.91	6908.10	24630.09	16246.57	35616.40	26626.75	100.00

Source: APEDA, 2014.

8.0 Important References

Ali, B.H., Blunden, G., Tanira, M.O. and Nemmar, A. 2008. Some phytochemical, pharmacological and toxicological properties of ginger (*Zingiber officinale* Roscoe): A review of recent research. *Fd. Chem. Toxicol.*, **46**: 409–420.

Al-Yahya, M.A., Rafatullah, S., Mossa, J.S. and Tariq, M. 1989. Gastroprotective activity of ginger (*Zinigiber officinale* Rosc.) in albino rats. *American J. Chinese Med.*, **17**: 51-56.

Anonymous. 1995. *Recommended international code of practice for packaging and transport of tropical fresh fruit and vegetables (CAC/RCP 44-1995)*, Codex Alimentarius, World Health Organization, Geneva and, Food and Agriculture Organization of the United Nations, Rome.

Anonymous. 2004. *Ginger -postharvest care and market preparation*, Bull. No. 23. Ministry of Fisheries, Crops and Livestock; National Agricultural Research Institute and New Guyana Marketing Corporation, Georgetown, Guyana.

Anonymous. 2005. *Codex standard for ginger (CODEX STAN 218-1999)*, Codex Alimentarius, World Health Organization, Geneva and, Food and Agriculture Organization of the United Nations, Rome.

Anonymous. 2005. *Herbs, spices and essential oils: Postharvest operations in developing countries*, United Nations Industrial Development Organization, Vienna, Austria and, Food and Agriculture Organization of the United nations, Rome, Italy.

Anonymous. 2014. *Basic Report 11216, Ginger root, raw*. USDA National Nutrient Database, The National Agricultural Library.

Anonymous. 2014. Ginger. http://agriexchange.apeda.gov.in/Market per cent 20Profile/one/GINGER.aspx, APEDA AgriXchange, browsed on July 15.

Anonymous. 2014. State/spice wise area and production of spices in India. http://www.indianspices.com/pdf/state_prd.pdf, Spices Board of India, Cochin, browsed on July 26.

ASTA. 2007. *ASTA cleanliness specifications for spices, seeds, and herbs (Foreign and domestically produced)*, American Spice Trade Association, Washington, DC 20036, USA.

Azam-Ali, S. 2008. *Ginger processing*, Practical Action, The Schumacher Centre, Bourton-on-Dunsmore, Rugby, Warwickshire, CV23 9QZ, United Kingdom.

Balakrishnan, K.V. 2005. Postharvest and industrial processing of ginger. **In**: *Ginger The Genus Zingiber*, P.N. Ravindran and K.N. Babu (Eds.). CRC Press, Washington, D.C., pp. 391-434.

Baranowski, J.D. 1986. Changes in solids, oleoresin, and (6)-gingerol content of ginger during growth in Hawaii. *Hort Sci.*, **21**: 140-146.

BIS. 2008. Indian Standard: Spices and condiments - Ginger, whole and ground - Specification. https://law.resource.org/pub/in/bis/S06/is.1908.2008.html, Bureau of Indian Standards, New Delhi, India, browsed on July 22, 2014.

Chang, J.S., Wang, K.C., Yeh, C.F., Shieh, D.E. and Chiang, L.C. 2013. Fresh ginger (*Zingiber officinale*) has anti-viral activity against human respiratory syncytial virus in human respiratory tract cell lines. *J. Ethnopharcol.*, **145**: 146-151.

Davis, C. and Wood, A. 2014. A literature review of the medicinal properties of ginger (*Zingiber officinale* Roscoe). **In**: *New Value-adding Opportunities for Natural Therapeutic Products in the Australian Ginger Industry*, C. Davis and J. Herse (Eds.). Horticulture Australia Ltd, Sydney. NSW 2000, Australia, pp. 59-113.

Dohroo, N.P. 2001. Etiology and management of storage rot of ginger in Himachal Pradesh. *Indian Phytopathol.*, **54**: 49-54.

Dohroo, N.P. 2005. Diseases of ginger. **In**: *Ginger The Genus Zingiber*, P.N. Ravindran and K.N. Babu (Eds.). CRC Press, Washington, D.C., pp. 305-340.

Dohroo, N.P., Kansal, S. and Ahluwalia, N. 2012. *Status of soft rot of ginger (Zingiber officinale Roscoe)*, Dr. Y. S. Parmar University of Horticulture and Forestry, Nauni, Solan-173 230 (HP).

FAO. 2014. FAOSTAT. http://faostat3.fao.org/faostat-gateway/go/to/browse/Q/QC/E, Statistics Division, Food and Agriculture Organization of the United Nation, Rome 00153, Italy, browsed on July 24.

Gao, D. and Zhang, Y. 2010. Comparative antibacterial activities of extracts of dried ginger and processed ginger. *Pharmacog. J.*, **2**: 41-44.

Govindarajan, V.S. 1982. Ginger - chemistry, technology, and quality evaluation: Part 1. *Crit. Rev. Fd. Sci. Nutr.*, **17**: 1-96.

Govindarajan, V.S. 1982. Ginger - chemistry, technology, and quality evaluation: Part 2. *Crit. Rev. Fd. Sci. Nutr.*, **17**: 189-258.

Haware, M.P. and Joshi, L.K. 1974. Studies on soft rot of ginger from Madhya Pradesh. *Indian Phytopathol.*, **27**: 158-161.

Kader, A.A. 2002. Pre and postharvest factors affecting fresh produce quality, nutritional value, and implications for human health. Proceedings of the International Congress on Food Production and the Quality of Life, Sassari, Italy, September 4-8, 2000, **1**: 109-119.

Larsen, K., Ibrahim, H., Khaw, S.H. and Saw, L.G. 1999. *Gingers of Peninsular Malaysia and Singapore*, Natural History Publications (Borneo), Kota Kinabalu.

Mukherjee, P.K., Thomas, P. and Raghu, K. 2008. Shelf-life enhancement of fresh ginger rhizomes at ambient temperatures by combination of gamma-irradiation, biocontrol and closed polyethylene bag storage. *Ann. App. Biol.*, **127**: 375–384.

NHB. 2013. *Indian Horticulture Database - 2013*, R.K. Tiwari, N.C. Mistry, B. Singh and C.P. Gandhi (Eds.). National Hortulture Board, Ministry of Agriculture, Govt. of India, Gurgaon 122015.

Ogogi, T., Minami, M., Kuraishi, Y. and Satoh, M. 1992. Capsaicin-like effect of 6-shogol on substance P-containing primary afferents of rats, a possible mechanism of iots analegisic action. *Neuropharm.*, **31**: 1165-1169.

Okwuowulu, P.A. and Nnodu, E.C. 1988. Some effects of prestorage chemical treatments and age at harvesting on the storability of fresh ginger rhizomes (*Zingiber officinale* Roscoe). *Trop. Sci.*, **28**: 123-125.

Onyenekwe, P.C. 2000. Assessment of oleoresin and gingerol contents in gamma irradiated ginger rhizomes. *Nahrung*, **44**: 130-132.

Paull, R.E. and Chen, C.C. 2014. Ginger. http://www.ba.ars.usda.gov/hb66/067ginger.pdf, United States Department of Agriculture, Beltsville, MD 20705, USA, browsed on July 5.

Paull, R.E., Chen, N.J. and Goo, T.T.C. 1988. Compositional changes in ginger rhizomes during storage. *J. American Soc. Hort. Sci.*, **113**: 584-588.

Plotto, A. 2002. *Ginger - Postharvest operations*, Food and Agriculture Organization of the United Nations (FAO), AGST.

Premavalli, K.S. 2005. Ginger as a spice and flavourant. **In**: *Ginger The Genus Zingiber*, P.N. Ravindran and K.N. Babu (Eds.). CRC Press, Washington, D.C., pp. 509-526.

Pruthi, J.S. 1980. *Spices and Condiments: Chemistry, Microbiology and Technology*. Academic Press, New York.

Pruthi, J.S. 1992. Postharvest technology of spices: Pretreatments, curing, cleaning, grading and packing. *J. Spices Aromatic Crops*, **1**: 1-29.

Purseglove, J.W., Brown, *E.G.* and Green, C.L. 1981. *Spices*, **Vol. 1**. Longman, London and New York.

Rahman, H., Karuppaiyan, R., Kishore, K. and Denzongpa, R. 2009. Traditional practices of ginger cultivation in Northeast India. *Indian J. Trad. Know.*, **8**: 23-28.

Rai, S. and Anitha, G. 1999. Mother rhizome extraction of ginger (*Zingiber officinale* Roscoe) - an age old practice in Sikkim and Darjeeling hills. *Envir. Ecol.*, **4**: 910-912.

Ravindran, P.N. and Babu, K.N. 2005. *Ginger The Genus Zingiber*. CRC Press, Washington, D.C.

Ravindran, P.N. and Babu, K.N. 2005. Introduction. **In**: *Ginger The Genus Zingiber*, P.N. Ravindran and K.N. Babu (Eds.). CRC Press, Washington, D.C., pp. 1-14.

Saikia, S. 1999. *Assessment of Growth Performance and Characterization of Products of Some Ginger (Zingiber officinale Rosc.) cultivars*. Ph.D thesis submitted to Assam Agricultural University, Jorhat, Assam.

Sasikumar, B., Thankamani, C.K., Srinivasan, V., Devasahayam, S., Santhosh, J., Eapen, A.K. and John, T.Z. 2009. *Ginger (Extension Pamphlet)*, Spices Board, Ministry of Commerce and Industry, Cochin - 682 025.

Sharma, S.K. and Dohroo, N.P. 1991. Postharvest management of rhizome rot (*Fusarium oxysporum* f.sp. *zingiberi* Trujillo) of ginger through chemical and antagonist. *Indian Cocoa Arec. Spices J.*, **14**: 150-152

Shoji, N., Iwasa, A., Takemoto, Y.I. and Ohizumi, Y. 1982. Carditonic principles of ginger. *J. Pharma. Sci.,* **71**: 1174-1175.

Singh, H.B. 2013. *An Assessment of Pre and Postharvest Factors Affecting Quality of Ginger in The Export Chain - A Case Study of Salyan District and Nepalgunj City, Mid-Western Development Region, Nepal.* Master of Agricultural Production Chain Management thesis submitted to Van Hall Larenstein University of Applied Sciences, Wageningen, The Netherlands.

Soepadmo, E. 1976. Ginger plants. *Nat. Malaysia,* **1**: 32-39.

Srinivasan, K. and Sambaiah, K. 1991. The effect of spices on cholesterol 7-alpha-hydroxylase activity and on serum and hepatic cholesterol levels in the rats. *Int. J. Vit. Nutr. Res.,* **61**: 364-369.

Srivastava, K.C. and Mustafa, T. 1989. Ginger (*Zingiber officinale* Rosc.) and rheumatic disorders. *Med. Hypoth.,*. **29**: 25-28.

Stoilova, I., Krastanov, A., Stoyanova, A., Denev, P. and Gargova, S. 2007. Antioxidant activity of a ginger extract (*Zingiber officinale*). *Fd. Chem.,* **102**: 764-770.

Sutarno, H., Hadad, E.A. and Brink, M. 1999. *Zingiber officinale* Roscoe. **In**: *Plant Resources of South-East Asia, No 13 - Spices*, C.C. de-Guzman and J.S. Siemonsma (Eds.). Backhuys Publishers, Leiden, The Netherlands, pp. 239-244.

Teakle, D.S. 1965. Fusarium rhizome rot of ginger in Queensland. *Queensland J. Agric. Anim. Sci.,* **22**: 263-272.

Trikarunasawat, C. 2008. *Biological Control of Rhizome Rot of Ginger during Storage by Antagonistic Microorganisms and Medicinal Plant Extracts.* Doctor of Philosophy (Plant Pathology) thesis submitted to Kasetsart University, Bangkok 10900, Thailand.

Trujillo, E.E. 1963. Fusarium yellows and rhizome rot of common ginger. *Phytopathol.,* **53**: 1370-1371

Wang, H. and Ng, T.B. 2005. An antifungal protein from ginger rhizomes. *Biochem. Biophys. Res. Comm.,* **336**: 100-104.

Weiss, E.A. 1997. *Essential Oil Crops.* CAB International Publishing, Oxon, UK.

Weiss, E.A. 2002. *Spice Crops.* CAB International Publishing, Oxon, UK.

Wu, J.J. and Yang, J.S. 1994. Effect of irradiation on the volatile compounds of ginger rhizome (*Zingiber officinale* Roscoe). *J. Agric. Fd. Chem.,* **42**: 2574-2577.

Chapter 32
Oil Palm

S. Arulraj

Director, ICAR-Directorate of Oil Palm Research,
Pedavegi – 534 450, Andhra Pradesh, India
E-mail: dopr2009@gmail.com

1.0 Introduction

Oil palm (*Elaeis guineensis* Jacq.) is a native of West Africa and is grown extensively in South-East Asian countries (Malaysia, Indonesia and Papua New Guinea), African countries (Nigeria, Cote de'Ivoire, Ghana, Liberia, Sierra Leone, Cameroon, Republic of Congo and Zaire) and South American countries (Costa Rica, Panama, Columbia, British Guyana, Peru, Ecuador, Venezuela and Brazil). Malaysia, Indonesia and Nigeria are the leading producers of oil palm in the world. Oil palm is a humid crop which requires evenly distributed rainfall of 150mm/ month or 1800 mm/annum. As rainfall distribution is not even and adequate in India, the crop could be grown only with assured irrigation source.

Oil palm is known to be the highest edible oil yielding perennial crop that produces two distinct oils, *i.e.*, palm oil and palm kernel oil, which have culinary and industrial uses. Palm oil is derived from the fleshy mesocarp of the fruit, which contains about 45-55 per cent of oil. The palm kernel oil, obtained from the kernel of stony seed, is a potential source of lauric oil. Oil palm has greater advantage in terms of productivity that is much higher than that other oil yielding crops. It produces 4 to 6 tonnes of crude palm oil per ha annually and 0.4 to 0.6 tonnes of palm kernel oil from 4th to 30th year of its productive life span.

Oil palm is a crop of the present and future vegetable oil economy of the world as well as for India. Palm oil has good consumer acceptance as a cooking medium because of its price advantage. It is a good raw material for manufacturing oleo-chemicals used in the preparation of cosmetics, pharmaceuticals, nutraceuticals etc. Broadly, it could be mentioned that palm oil is a source for improving health and nutrition, value addition, waste utilization, eco-friendly, source of diversification, import substitution, co-generation and sustainability.

2.0 Status of Cultivation in India

Oil palm was first introduced to India at National Royal Botanical Gardens, Kolkata during the year 1886. The Maharashtra Association for Cultivation of Sciences (MACS), Pune later introduced African dura palms along canal bunds, home gardens and, to some extent, in forest lands near Pune during 1947 to 1959. Large scale planting of oil palm was launched from 1971 to 1984 in Kerala State by Plantation Corporation of Kerala Ltd. and Andaman Forest and Plantation Development Corporation Ltd., in Little Andaman Islands of Andaman and Nicobar Islands during 1976 to 1985. Oil palm has established as a successful crop in a number of states in the country and productivity levels of 6-8 tonnes of oil per ha could be achieved. The Technology Mission on Oilseeds and Pulses (TMOP) implemented by Government of India looks after development of oil palm in the country through Oil Palm Development Programme (OPDP) along with other nine annual oilseed crops.

Though Oil Palm Development Programme in the country is progressing well, area coverage is not taking place as per targets envisaged. As on 2013, various Expert Committees constituted by Ministry of Agriculture, Government of India have identified a total of 19.33 lakh hectares in 18 states of the country as suitable for oil palm cultivation. An area of 2.32 lakh ha only has been covered under oil palm upto 2012-13. (Table 1). Production of palm oil in India continues to be at a meagre level with respect to its actual requirement. The FFB yield obtained by progressive farmers of Andhra Pradesh and Karnataka, under optimum management conditions, is between

Table 1: Area Under Oil Palm Cultivation-2013

State	*Area Planted (ha)*
A & N	1593
Andhra Pradesh	142079
Chattisgarh	61
Goa	857
Gujarat	3234
Karnataka	30696
Kerala	5394
Maharashtra	1394
Mizoram	18016
Odisha	13119
Tamil Nadu	16074
Tripura	31
Total	**232548**

20 and 25 tonnes of FFB per ha per annum *i.e.* 4-5 tonnes of oil per ha per annum from fourth year onwards. High yields of 30-35 tonnes FFB/ha from seventh year onwards was also recorded in many plantations. One of the farmers in Karnataka could achieve a record annual mean yield of 53.20 tonnes of FFB per ha.

3.0 Major Commercial Varieties

Oil palm belongs to Palmae family of the order palmales. It is monoecious- male and female inflorescences are produced separately on the same palm in cycles. Oil palm fruit is a sessile drupe. The pericarp of the fruit consists of the outer exocarp (skin), mesocarp (or pulp rich in oil) and endocarp (hard stony shell surrounding kernel). The kernel has testa (skin), a solid endosperm (oil rich) and embryo. The fruit development takes place from about the 15th to 19th day after anthesis and oil formation takes place towards the end of maturity during which shell hardens and the embryo becomes viable. There are three types of oil palm fruits which impart characteristic feature to a palm as:

a) Dura

Dura palms have fruits with thick shell (2-8mm in thickness), low to medium mesocarp content (33-55 per cent) and are not commercially grown. These are used as female parent (mother palms) in developing tenera hybrid.

b) Pisifera

Pisifera palms are female sterile, shell-less and used as male (pollen) parent in development of tenera hybrids.

c) Tenera

This is the commercial planting material which is the cross between dura (female parent) and pisifera (male parent). Tenera fruits have thin shell (0.5-4.0mm) with medium to high mesocarp content (60-95 per cent). There is prominent fibre ring in its mesocarp.

4.0 Nature and Causes of Postharvest Losses

4.1 Preharvest factors

The notion that "palm oil is formed in the field not in the mill; the mills merely process to recover it" is significant that bunches harvested are ripe and at their optimal oil formation stage. The best bet for palm oil mills to achieve high oil extraction rate (OER) is to target for high degree of FFB ripeness of more than 90 per cent. Tight management and control on FFB harvesting, collection and handling is necessary.

Harvesting is among the most important activities of the whole production process of the oil palm because of its high cost (18 per cent of the total cost of the farm). Proper and timely harvesting of fruit bunches is an important operation which determines the quality and quantity of oil to a great extent. Harvest should be aimed at recovering the whole harvestable produce without loss of oil or loose fruits. The yield is commonly recorded in metric tonnes of fresh fruit bunches (FFB) per hectare per year, or metric tonnes of oil per hectare per year.

Harvest and postharvest operations in oil palm have considerable effect on the quality of oil. Presence of an enzyme lipase embedded in the membrane of fat globules in the fruit mesocarp causes accumulation of Free Fatty Acid (FFA) content of palm oil. Over-ripeness, improper handling, delayed processing etc. damage the delicate membrane, which triggers the enzyme activity and releases FFA. High FFA content makes the palm oil non-edible, enhances the rate of rancidity, fixes colour and increases refining losses. Processing the fruits within 24 hours is recommended to retain edible quality of raw palm oil. Harvesting at right stage, preventing excessive bruising of fruits during harvest and transportation and prompt processing are the key factors to overcome this deficiency. However, those who consume crude palm oil directly, like the Africans, the high FFA content is often ignored. Indian consumers prefer Refined, Bleached and De-odorized (RBD) palm oil and hence the build up of FFA has to be minimized during harvesting, handling and processing of the fruits.

4.2 Degree of Ripeness

The oil content in the mesocarp which is the major economic produce of oil palm is strongly related to fruit ripeness. In the early stages, the mesocarp has high water and carbohydrate content, and very little oil. As the fruit ripens, the oil content increases and then formation of free fatty acids (FFA) increase. As the fruit matures, the percentage of extractable oil increases to a certain degree, but at the same time, its quality decreases due to its acidity. The highest concentration of oil occurs during the final weeks of development of the fruit. A week before harvest, the oil content can increase to 80 per cent under irrigated conditions.

Oil is stored in the mesocarp cells in little vacuoles separated from the enzymes (Lipoxidases) by a delicate membrane, which when torn by overripeness or handling injury allows the two components (oil and enzyme) to mix leading to the formation of acids. The fruit adhered to the bunch has a FFA content of only 0.2-0.9 per cent, while in the oil that comes out after proper extraction in a mill usually reach upto 2 per cent. Internationally, the present permissible limit of FFA in food grade palm oil is 5 per cent. FFA content in the oil is affected by natural content of FFA in the fruit, action of the lipase in the fruit, action of microbial lipase and autocatalytic activity after processing.

4.3 Factors Affecting Ripeness

Optimum ripeness, measured as a function of fruit detachment, varies according to plant age, and environmental conditions. In some cases the genetic material largest amount of oil was found when 30 per cent of the fruits had detached from the bunch. While in other case increase in oil content in the mesocarp was found in detachments greater than 6 per cent (after cut and gathered). A faster rate of ripeness has been observed in the tenera types in the smaller bunches and in the bunches of younger palms. Besides, those bunches which develop during summer (those pollinated at the end of the rainy season) ripe faster (162 days) than those bunches that develop during rainy periods (178 days).

An increase in oil/fresh mesocarp content was found when the number of fruits per bunch is less. According to this observation, bunches with more fruit (largest bunches) could be cut with a lower degree of ripeness. It is believed that bunches with less fruit are always associated with higher oil/ mesocarp content regardless of their ripeness degree.

A report which studied the oil accumulation rate and oil weight per bunch of palms of different ages found a linear increase in both. However, the increase in the rate of oil accumulation was higher than the oil weight per bunch because of water loss. According to this information, an increase in oil percentage in stored bunches is a false indicator, since it is due to moisture loss rather than oil synthesis.

4.4 Fruit Ripeness and OER

The oil extraction rate (OER) is the actual oil production in the mill and this depends on the inherent oil/bunch ratio, harvesting standards and practices in the plantation and the efficiency in the mill. Modern mills do not vary much in efficiency and hence the major contribution to this variation must be originating from harvest practices. It was reported in 2003 that Malaysian industry was losing US$ 250 million per year because of the low OER which in turn was attributed to the prevailing harvesting standards and procedures and suggested that the loose fruit collection should be given high priority. While 20 per cent OER was considered normal in Malaysia, many of the mills in Indonesia are obtaining OERs of around 24 per cent. One of the important causes suggested is low standards of harvesting, long intervals between harvesting rounds, harvesting of under-ripe fruits by harvesting too early or loss of high oil containing fruits because of too late and careless harvesting. With careless harvesting, many black bunches (unripe bunches) will be harvested leading to low OER. The oil extraction rate of the bunch is low during the first two or three years, but it reaches maximum values when the bunch weighs more than 4 kg on an average.

In India, the oil recovery rate (OER) is reported as 15 to17 percent at present. This results in a recovery gap of at least 4 percent which is equivalent to 11000 tonnes of crude palm oil (CPO), based on present statistics. Hence, correct maturity and grading are important in deciding the OER and thereby profitability of the crop. By following proper harvesting standards and grading, more oil could be produced (recovered) from the existing plantations.

5.0 Postharvest Management

5.1 Harvest Time

The commercial oil palm variety tenera usually starts producing fruit bunches when the plant is 2-3 year old. To ensure better growth of palm and also have good yield during the first harvest, it is suggested to have ablation *i.e.* removal of early bunches. Flowering starts from 14th to 18th month after planting. Ablation should be started immediately after the appearance of inflorescences on the palms upto six months before the expected start of harvest. Ablation increases the storage of carbohydrate in the trunk which acts as reserve and gives more uniform yield. Some studies show that root development was encouraged by ablation thus giving greater drought tolerance to the plant.

5.2 Maturity Standards

The oil palm farmers have to be fully aware of the maturity standards and proper handling of the fruit bunches to ensure the quality of palm oil. The most common type of fruit is deep violet to black at the apex and pale greenish yellow at the base before ripening. Apex of the fruit turns deep red and golden yellow at the base as the fruit ripens. All the fruits in a bunch do not ripen simultaneously. As and when an individual fruit fully ripens, it becomes easily detachable from the bunch. The FFA content of palm oil increases as the ratio of detached fruit to total fruit on the bunch increases. The main aim of timely harvesting is therefore to obtain FFB, which will give good quality oil. Therefore the frequency of harvesting in an area has to be kept sufficiently high to eliminate over-ripening.

Bunch ripeness is identified by detachment of the fruit, change in colour and change in fruit texture. Granting that the fruits of a bunch do not ripe all at once, the way to obtain the maximum amount of oil is harvesting more frequently those bunches with an optimum ripeness degree; However, since this is not economically feasible, the harvest could be done at regular intervals. The bunches harvested in every round are the ones that contain a greater number of fruits with optimum ripeness and fewer under-ripe or over-ripe fruits. To determine the optimum degree of ripeness of the bunch based on fruit detachment, the following criteria are basically followed:

i) Number of fruits that have fallen before cutting the bunch: 10 or more fruits detached or easily removable for young palms, and 5 for adult palms.

ii) Number of fruits detached according to bunch weight counted after the bunch has been cut: 5 or more fruits/kg of bunch.

iii) Amount of fruit detachment per bunch: fruit detachment is 25 per cent or more of the bunch visible surface. Any of the above criteria recommended must be flexibly applied. Aspects such as harvesting frequency, climatic conditions and type of fruit must be considered in an efficient harvesting system.

5.3 Harvesting Frequency

Oil synthesis and free fatty acid formation occur during the ripening process and hence harvesting rounds should be carried out as frequently as possible in order to reduce the number of over-ripe bunches to be harvested, which otherwise lead to a high degree of fruit detachment and increased acidity of oil.

Harvesting must take place in such a way that a bunch which is almost ready for one cycle will not be overripe for the following cycle. In low production periods, harvesting rounds can be made with less frequency, but the ripeness degree with which the bunches are to be cut should be less. This means that the longer the interval, the less is the fruit detachment accepted to cut the bunch (lower cutting order). According to current practices, harvesting rounds every 7-14 days have given good results. Factors determining harvest intervals basically refer to season, age of palms, labour availability, workers' skill and knowledge, transportation capability and processing capacity of the mill.

5.4 Harvesting Tools

In young palms, a 6-9 cm wide chisel attached to the tip of a 1.2-1.5 m long stick of wood or light hollow metal pipe with a handle is used for harvesting the bunches. With this kind of tool, the bunches can be cut without damaging the petiole of the leaves, and since the bunch is cut from a certain distance, workers are not hurt by the spines of the palm frond. When the harvest is started, it is advisable to cut down the bunch trying not to eliminate the leaf that supports it. Leaves are cut only to facilitate access to the bunch. The general rule for young palms is to avoid cutting down any green fronds unless it is unavoidable. During this task, it is important not to injure the trunk and the petiole of neighboring leaves. At the same time, rotten bunches (badly pollinated, diseased or failed) as and when found, must be eliminated in order to clean the plant. Use of the narrow chisel is usually continued until the palm reaches a height of 1m above the ground. The narrow chisel is substituted by a wider one (14 cm wide) attached to a longer stick or metal pipe which should be longer as the palm grows (1-3 m long). Wide chisels used for cutting bunches was found suitable for palms not higher than 3-4 m in height. Even though the bunches are cut from a distance, it is always advisable to use long leather gloves and cap, to give the harvesters protection for their hands and forearms.

In taller plants, bunches are cut by using pole and sickle/knife. A curved knife is attached to the end of a long thin and light pole which could be made of bamboo, aluminum or any other light and strong material. The knife can be fixed at the end of the pole with steel wire or with screws. There are different sickles available for the purpose *viz.* Malaysian sickle, modified Malaysian sickle, Kerala sickle, Andhra Pradesh sickle, Tamil Nadu sickle etc. DOPR1 and DOPR5 sickels are found to be ideal. Cutting bunches with pole and sickle requires skilled workers. A good cut depends not so much on strength but on the position/angle of the knife blade. There are ripe bunches whose base is difficult to find space to insert the tip of the knife to make the cut. Under these circumstances, the worker must find the right standing location to cut the bunch without hurting the young leaves of the palm. Cutting of subtending leaf is accepatable in case of adult palms since; the leaf area available is still adequate after cutting a few fronds.

In those fields where the height of the palm is uneven, short poles cannot be used with tall palms, and if a long pole is used for harvesting a short palm, the worker has to stand far from the palm and the task becomes more difficult. In these particular cases, one pole is used inside another (telescopic type). The inside pole can be used as an extension when the height of the palm requires it. When these instruments are transported, the blades must be covered properly in order to avoid accidents.

5.5 Harvest Management

Once the harvest interval has been decided, the area to be harvested every day must be determined. For this purpose, the total area is divided among the number of days of the planned interval. Then the area to be harvested is divided by the workers' harvesting capacity. In this way, number of workers needed to harvest each section within the desired interval can be determined. The aspects determining area, size, or number of rows to be harvested are i. Height of the palm: The taller the palm, the greater skill required to handle the cutting tools. ii. Number of bunches per area: When there are few bunches per area unit, the harvester must walk longer distances (more rows) to complete his task. iii. Average bunch weight: More effort is required to carry and transport heavier bunches. iv. Carrying distance: When collection centers are close to each other, bunch carrying is easier. v. Field conditions: Variations in topography, weed growth, poor drainage etc could affect the workers' efficiency.

5.6 Sorting and Grading

The primary objectives of oil palm bunch grading are To improve the quality of fresh fruit bunches (FFB) received at mills To improve the quality of palm oil produced; To improve the efficiency of oil and kernel extraction rates. To ensure that farmers and processors obtain a fair deal from transactions and in India, as of date, there is no grading practiced for oil palm FFB. Research work on this line is in progress considering the local features. However, owing to its importance in achieving better productivity and quality of oil, details of the grading system, as developed in Malaysia, are presented in Table 2.

Table 2: Palm Oil Registration and Licensing Authority (PORLA) Classification of Oil Palm FFB (at the time of inspection at mill)

Bunch Grades	*Description*
Dirty Bunch	With more than half of its surface covered with mud, sand, other dirt particles and mixed with stone or other foreign matters.
Diseased Bunch	Has more than fifty percent (50 per cent) parthenocarpic fruits and is not normal in terms of its size or its density.
Dura Bunch	With shell thickness of 2-8 mm, Ratio of shell to fruit of 25-50 per cent, Ratio of mesocarp to fruit of 20-60 per cent, Ratio of kernel to fruit of 4-20 per cent and with no fibre ring around the shell
Empty Bunch	Has more than ninety percent (90 per cent) of detached fruitlets.
Long Stalk Bunch	Has a stalk of more than 5 cm in length (measured from the lowest level of the bunch stalk).
Loose Fruit	Fruit detached from a fresh fruit bunch because of ripeness and is reddish orange in colour.
Old Bunch	Has been harvested and left at the field before being sent to the mill. The fruits still remaining on the bunch are dry and brownish black in colour. The stalk is also dry, soft, fibrous and blackish in colour.
Overripe Bunch	Fruits are darkish red coloured and has more than fifty percent (50 per cent) of detached fruitlets but with at least ten percent (10 per cent) of the fruits still attached to the bunch.
Pest Damaged Bunch	With more than thirty percent (30 per cent) of its fruits damaged by pest attack such as rats, birds etc.
Ripe Bunch	Reddish orange in colour, outer layer fruit mesocarp is orange in colour, has at least 10 fresh sockets of detached fruitlets and more than fifty percent (50 per cent) of the fruits still attached to the bunch.
Rotten Bunch	Partly or wholly and together with its loose fruits, has turned blackish in colour, rotten and mouldy.

Contd...

Table 2–*Contd...*

Bunch Grades	*Description*
Small Bunch	Has small fruits and weighs less than 2.3 kg.
Under ripe Bunch	Reddish orange or purplish red in colour, outer layer fruits mesocarp yellowish orange in colour and has less than 10 fresh sockets of detached fruitlets.
Unfresh bunch	Has been harvested and left at the field for more than 48 hours before being sent to the mill. The whole fruit or part of it together with its stalk has dried out. Normally, this type of bunch is dry and blackish in colour.
Unripe Bunch	Black or purplish black in colour, outer layer fruits mesocarp is yellowish in colour, does not have any fresh sockets of detached fruitlets and sockets (if any) on the bunch are not due to normal ripening process.
Wet Bunch	Consignment of fresh fruit bunches (FFB) which has excessive free water.

Palm Oil Registration and Licensing Authority of Malaysia under the Ministry of Primary Industries, has developed a Fresh Fruit Bunch Grading Manual during the year 1995 which has been given hereunder. The manual serves as a practical guide for grading of oil palm bunches based on various quality parameters.

Since the basic factor determining Oil Extraction Rate (OER) is the quality of FFB, any step to improve the OER should begin from the plantations. In order to ensure quality control in plantations, there should be a mechanism in the processing units to check the FFB quality. Quality control in processing units would act as a corrective measure for the farmers to improve their produce. Fresh Fruit Bunch grading is such a process whereby FFBs are assessed and classified according to the degree of ripeness and quality. Palm oil mills in Malaysia follow such grading systems.

Successful implementation of grading requires well-defined grades according to quality. Different FFB quality grades defined by Palm Oil Registration and Licensing Authority (PORLA), suitable for Malaysian conditions.

Trained grading staff, having capability and experience in the grading of fresh fruit bunches (FFB), are appointed by the mills to ensure bunch quality. Well-defined sampling methods and step by step grading procedure is also in force to facilitate impartial grading. The minimum sample size of each consignment to be graded is based on the net weight of the consignment. In practice, it is not possible to obtain hundred percent good quality bunches and hence a reasonably good quality consignment constitutes the combination of bunch quality (Table 3).

In order to facilitate payment according to basic oil extraction rate, standard extraction rates based on age of palm and bunch weight are established for different climatic regions. The graded oil and kernel extraction rate that can be considered for payment is calculated by subtracting penalty rate from the basic extraction rate. The graded extraction rate is recorded in the monthly grading summary for the purpose of payment to the suppliers.

Table 3: Bunch Quality in Oilpalm

Bunch Category	*Limit*
Dirty Bunch	0 per cent
Dura bunch	0 per cent
Empty Bunch	0 per cent
Long stalk Bunch	≤ 5 per cent
Old Bunch	0 per cent
Ripe Bunch	≥ 90 per cent
Rotten Bunch	0 per cent
Under ripe Bunch	≤ 10 per cent
Unripe Bunch	0 per cent
Wet Bunch	0 per cent

Case studies on the factors affecting OER in palm oil mills of Malaysia report that OER of above 19 per cent is possible by improving the FFB quality by observing the grading and quality control procedures, keeping other factors constant. One of the mills which reported a high percentage of unripe (5.23 per cent) and under ripe bunches (34.42 per cent) was capable to give only 17.10 per cent OER could improve the OER above 18 per cent by following 100 per cent FFB grading, rejecting black and rotten bunches and imposing penalty for long stalks by weight deduction.

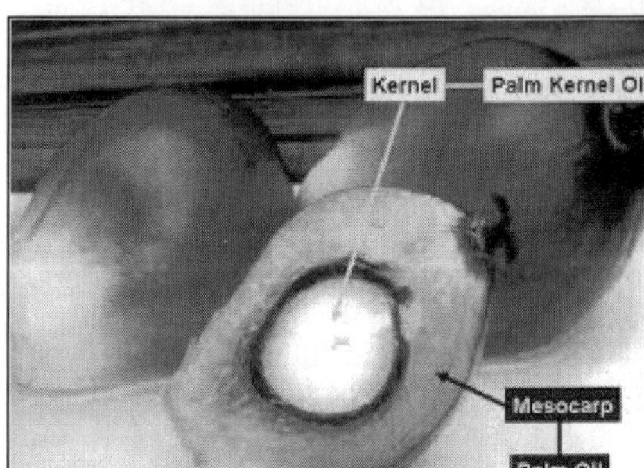

Figure 1: Kernel of Oil Palm

In India, as far as grading and quality control is concerned oil palm cultivation being in the infant stage, the Government and industry have not given much to this aspect and grading system is not implemented so far in Indian palm oil mills. The kernel of oil palm is depicted in Figure 1. However, recently, after the CACP recommendation for Price Fixation Formula for oil palm was recommended by Government of India, attempts were made by a few Processing Units for informal control on the quality of bunches and both farmers and Processors realized the benefits within a few months. Keeping area expansion in view, as the crop is gaining momentum, it was felt that rigid grading would create adverse impact on area expansion. Accordingly development officials appointed by processors as well as Scientists of Directorate of Oil Palm Research are making efforts to educate farmers about the importance of quality control and the ways to achieve good quality bunches. But harvesting of under ripened and unripened bunches due to reasons other than the knowledge level of farmers could not be prevented and has resulted in less OER in the past years. Grading of FFB in the processing mills would act a corrective measure to improve OER and subsequently the FFB price realized by the farmers.

The Malaysian (PORLA) grading procedure and classification of bunches may not be suitable for Indian conditions due to obvious variation in maturity standards, bunch characteristics, FFB procurement methods etc. Hence a modified procedure suitable for Indian conditions is needed for successful implementation of such initiatives.

Fresh Fruit Bunches (FFB) are to be processed within 24 hours after harvesting to retain edible quality of crude palm oil. At present, India has 23 palm oil processing mills and the capacity ranges from 4 tonnes/hr to 60 tonnes/hr. Overall processing capacity of the mills is 266.3 tonnes/hr.

5.7 Storage

Oil palm fresh fruit bunches (FFB) cannot be stored after harvesting, but should be processed within 24 hours. Studies conducted at DOPR on the effect of low temperature on FFA content in the oil after harvesting of FFB indicated that the FFA content increase was steady upto seven days and hence there is no possibility of storing FFB at lower temperature before processing.

5.8 Transportation

In India, the Technology Mission on Oilseeds and Pulses constituted by the Government of India after the development of oil palm in the country was looking National Mission on Oil seeds and Oil palm (NMODP) was launched in April. 2014 for improving vegetable oil production in the country. The State Governments of respective states have identified Entrepreneurs for establishing oil palm plantations and creating processing facilities in the states. The entrepreneurs are involved in the oil palm development programme from raising of the seedlings to providing inputs and guidance to the farmers and setting up collection centres for collection of oil palm fresh fruit bunches and processing. Oil palm fresh fruit bunches, after harvesting, are transported by the farmers to the collection centres established by the oil palm processing units nearer to the farmers fields or to the respective processing mill, either by tractors or bullock carts.

6.0 Domestic Marketing

The Oil Palm Act, implemented by various state governments for the benefit of oil palm farmers as well as processors, facilitates the buy back of the oil palm FFB by the Oil palm processors. Presently, the State Governments are deciding the prices FFB as per the recommendation of the price fixation committee under the Chairmanship of Secretary (Agriculture/Horticulture) of the concerned State Government – based on the price fixation formula developed by CACP and recommended by Government of India. The price of FFB are influenced by the international and national price of palm oil.

7.0 Conclusion

With the present level of rapid increase in the per capita consumption of vegetable oils in the country, it is expected that the estimated population of 1620 million in

2050 may need 40.5 million tonnes of vegetable oils at the estimated per capita consumption of 25.00 kg. Though it would be ideal, if the country could become self-reliant in the vegetable oil production sector, it would not be possible for us to spare the required land in view of demand for additional area for different crops as well as enterprises including urbanization. In the recent report on the potential area for oil palm cultivation in India, DOPR recommended that an area of 1.93 million ha from 18 States could be utilized for oil palm cultivation. With intensive research efforts and active implementation of Oil Palm Development Programme, it is estimated that the oil palm productivity levels in the country could be enhanced to 7 tonnes of oil per ha, thus producing a total quantity of about 13.51 million tonnes of palm oil by 2050. Thus, the oil palm could contribute towards the vegetable oil requirements of the country by 33.35 per cent.

The agricultural paradigm is already undergoing a shift with focus from cereal production to diversified farming. Oil palm besides improving biological productivity and nutritional standards also has enormous scope for enhancing profitability. Past investment in oil palm research has been rewarding in terms of increased production and productivity of oil palm. However, challenges confronting oil palm sector are still many. While the edible oil demand is increasing But the production has to be competitive both in terms of quality and price. Improving the efficiency of oil extraction process and reducing the losses are also equally important from the point of view of boosting the oil production. Thus, the development of improved hybrids with high productivity, quality characteristics, resistance to pests and diseases and tolerant to abiotic stresses need prime attention. The technologies must improve the efficiency of water and nutrients and reduce variability in yield and quality and also reduce postharvest losses. Efforts will also be needed to ensure self sufficiency in production of quality hybrid seed and planting material. Development and popularization of Good Agricultural Practices in preharvest and postharvest management of oil palm would contribute significantly towards improving the system efficiency in oil palm sector.

8.0 Important References

Carrere, R. 2010. Oil palm in Africa; Past, present and future scenario. World rain forest movement *http://wrm.org.uy/countries/Africa/Oil-Palm-inAfrica.pdf.*

Jelani, A.R., Hitam, A., Shuib, A.R., Jamak, J., Noor, M. and Gono, Y. 2005. Motorised Mechanical Cutter for Efficient Oil Palm Harvesting. *Proceedings of the PIPOC 2005 International Palm Oil Congress (Agriculture, Biotechnology and Sustainability).* pp. 1189-1193.

Malike, F.A, N. Abdullah and Ahmad, N., 2011. Evaluation of Harvest Index in the MPOB germplasm collections. pp 172-175. **In:** *Proceedings of Agriculture, Biotechnology and Sustainability Conference, PIPOC, 2011,* 15-17, November, KLCC, Malaysia.

Murugesan, P., Bijimol, G., and Gopakumar, S. 2009. Nut Component Analysis of Exotic and Indigenous Sources of Oil Palm (*Elaeis guineensis,* Jacq.) Planting Materials. *Int. J. of Oil Palm,* **6(1):** 19-21.

Shuib, A.R., Jelani, A.R., Jahis, S., Deraman, M.S., Khalid, M.R., Hitam, A., Basiron, Y. and Wahid, M.B. 2004. Development of a Machine for Harvesting Tall Palms, *MPOB TT No. 217.*

Webb, B.A.1976. The development of suitable fruit harvesting system to improve the efficiency of oil palm production. *Proceedings of Agriculture Oil Palm Conference, Kuala Lumpur:* pp 741-752.

Chapter 33
Onion

R.P. Gupta

Director

H.P. Sharma

Scientist

National Horticultural Research and Development Foundation,
Nashik – 422 003, Maharashtra, India, India

E-mail: nashik@nhrdf.com

1.0 Introduction

The onion (*Allium cepa* L.) is the most important commercial vegetable grown and consumed almost all over the world. Onion bulbs have been used as a food source since ancient times. Traces of onion remains were found alongside date stones and fig remains from Bronze Age that dates back to 5000 BC. However, it is not clear if these were cultivated onions. Although there is no clear cut evidence it is believed to have been originated in the central Asia as evidenced by onion traces being found in the eye sockets of Egyptian burials.

Onion is grown under a wide range of climatic conditions, *i.e.* temperate, tropical and subtropical. It grows well under mild climate without extreme heat or cold or excessive rainfall. Its growth, bulb development and bolting or flowerings are influenced by photoperiod and temperature. It does not thrive well when average rainfall exceeds 75-100 mm during monsoon period. The ideal temperature for its vegetative growth is 12.8-23.0 °C. Warm temperature favours good development of bulbs. In India onion varieties grown in plains are of short day type, requiring 10-11.5 hours day length, while long day types requires 13-14 hours day length. These long day types are cultivated in hills. Sudden rise in temperature in *rabi* season results in development of small sized bulbs that matures early. The long day varieties do not develop bulb under short day conditions, whereas short day varieties if planted under long day conditions develop bulb. Onion has been reported to contain various organosulfur compounds that have antibacterial and anti-carcinogenic properties and are rich in flavonoid like quercetina valuable natural source of antioxidant. Carbohydrates in onion constitute about 80 per cent of dry matter and the major non-structural carbohydrate of onion bulb is fructo-oligosaccharides, well known as fructan, followed by glucose, fructose, and sucrose. Studies conducted at several places revealed that the postharvest losses of onion range from 15-40 per cent. Among the different components of postharvest losses the decay losses contribute to around 8-10 per cent whereas the losses due to sprouting and desiccation ranges from 5-8 per cent and 20-25 per cent, respectively, depending upon the period of storage.

2.0 Status of Cultivation

2.1 Global Scenario

Onion is grown in the world (Table 1) covering an area of 42.04 lakh hectares with a total production of 828.52 lakh tons and productivity of 19.71 tons/ha.China and India are leading onion growing countries followed by U.S.A., Iran (Islamic Republic of), Russian Federation, Egypt, Turkey, Pakistan, Brazil, Netherlands,

Table1: Global Scenario of Major Onion Producing Countries (2012)

Area Name	*Area (Lakh Ha)*	*Production (Lakh)*	*Productivity (T/ha)*
Algeria	0.46	11.83	25.57
Bangladesh	1.36	11.59	8.54
Brazil	0.61	15.19	24.93
China	10.25	226.00	22.05
Egypt	0.60	20.25	33.75
India	9.59	163.09	17.01
Iran (Islamic Republic of)	0.71	22.60	31.83
Japan	0.25	10.97	44.06
Mexico	0.43	12.39	29.08
Myanmar	0.72	11.40	15.83
Netherlands	0.27	13.53	49.68
Nigeria	1.95	13.50	6.92
Pakistan	1.30	16.92	13.05
Republic of Korea	0.21	11.96	57.03
Russian Federation	0.92	20.81	22.59
Spain	0.23	11.87	52.07
Turkey	0.63	18.19	28.87
U.S.A.	0.60	32.77	54.62
Ukraine	0.64	11.41	17.80

Source: FAO Website.

Nigeria, Mexico, Republic of Korea, Spain, Algeria, Bangladesh, Ukraine, Myanmar, Japan. The productivity of onion is maximum (57.03tons/ha) in Republic of Korea, followed by USA (54.62tons/ha), Spain (52.07tons/ha) and Netherlands (49.68tons/ha) while the productivity in India is quite low (17.01tons/ha).

2.2 National Scenario

Onion is grown widely in different parts of the country. It is important foreign exchange earning crop to the country. India produces around 168.13 lakh tons of onion annually ranking second in the world. The production has increased more than five times during past three decades. The reason for increase in production is mainly due to increase in area under onion cultivation from 1.94 lakh ha during 1974-75 to 10.52 lakh ha in 2012-13. India possesses many innate advantages over other onion producing countries due to its large genetic base, varied soil and climatic conditions and skilled human resources. However, the productivity is still low. Poor soil fertility, use of low level of inputs like manures, fertilizers, crop protection chemicals, high labour cost and crop loss due to diseases, lack of varieties resistant to pests and diseases are the major reasons for low productivity. Among the different onion growing states, Maharashtra ranks first in area and production, followed by Karnataka, Madhya Pradesh and Andhra Pradesh. Maharashtra has highest share both in area (24.73 per cent) and production (27.72 per cent).However, the average productivity was found to be the highest in Gujarat (24.42 tons/ha), followed by M.P., J&K and Punjab states (Table 2).

2.3 State-wise Status of Onion Production in India

Two to four crops of onions are now taken in various parts of the country. Fresh onion starts coming from July onwards in the markets from Kurnool in the State of Andhra Pradesh, Tirpur in Tamilnadu,Chitradurga and Chellikere area of Karnataka and continues upto April-May.In Maharashtra State, fresh onion starts coming in market in September-October (Satara) and continues till April-May from Nasik, Ahmednagar, Pune and other district. Gujarat State produces three crops *i.e.* Kharif(November-January), late kharif(February-March) and rabi(April) in Bhavnagar, Amreli, Rajkot, Junagarh and Jamnagar districts. Kharifonion production although not common in north Indiais now taken up in some pockets of Rajasthan (Alwar, Dausa, Bharatpur and Ajmer), Haryana (Rewari, Mewat and Gurgaon districts) and Punjab (Patiala) states. In Northern, Eastern and Central parts of India, *rabi* onion is commonly cultivated which is available in April-May. Multiplier onion (shallot) is grown in Tamilnadu and Odisha states. The quality of multiplier (shallot) onion produced in Odisha is better. Small pickling onion is produced in Andhra Pradesh near Mydukur in Cuddapah district during kharifseason and in Karnataka near Chickballapur and Bangalore in all the three seasons exclusively for exportpurpose. The quality of small onion from Karnataka is better due to prevalence of mild climate almost round the year. The availability of onion in India from different locations is mentioned in Figure 1.

3.0 Major Commercial Varieties

A large number of onion varieties for cultivation in different regions and also in different seasons have been developed in India. The high quality seed of all onion

Table 2: Area, Production and Productivity of Onion in India (2012-2013)

State	*Area ('000 ha)*	*Production (in '000 MT)*	*Productivity (Ton/ha)*
Andhra Pradesh	86.67	1,560.06	18.00
Assam	8.53	30.90	3.62
Bihar	53.02	1,107.84	20.89
Chhattisgarh	17.95	269.28	15.00
Delhi	0.81	15.50	19.14
Gujarat	28.85	704.38	24.42
Haryana	27.80	604.47	21.74
Himachal Pradesh	2.27	39.36	17.34
Jammu and Kashmir	2.85	65.27	22.90
Jharkhand	17.21	322.15	18.72
Karnataka	159.60	2,395.90	15.01
Madhya Pradesh	111.73	2,691.00	24.08
Maharastra	260.00	4,660.00	17.92
Meghalaya	0.47	3.95	8.40
Mizoram	0.43	3.96	9.21
Nagaland	0.65	6.00	9.23
Orissa	34.92	419.09	12.00
Pondicherry	0.01	0.09	9.00
Punjab	8.26	183.65	22.23
Rajasthan	139.05	476.21	3.42
Sikkim	0.32	1.71	5.34
Tamil Nadu	37.70	429.72	11.40
Uttar Pradesh	26.63	474.01	17.80
Uttarakhand	3.82	39.40	10.31
West Bengal	22.00	309.10	14.05
Total	**1,051.55**	**16,813.00**	

Source: National Horticultural Board (http://nhb.gov.in/area per cent 20_production.html.

varieties developed by NHRDF is being produced and distributed and popularized by NHRDF all over the country. These varieties of onion have been widely accepted and commercially grown by the farmers in the country as well. As result of which there is considerable improvement in production, productivity and availability of onion for domestic and export markets.The list of onion varieties developed by different agencies in the country is given in Table 3.

4.0 Nature and Causes of Postharvest Losses

In onion many operations are performed in preharvest and postharvest stages for quality retention and reduction of postharvest losses. Cumulative losses at different stages of handling of onion range from 35 to 50 per cent. The major loss in onion after

	Jan	Feb	Mar	Apr	May	Jun	Jul	Aug	Sep	Oct	Nov	Dec
Andhra Pradesh		■	■	■			■	■	■	■	■	
Bihar					■	■					■	■
Gujarat	■	■	■	■	■					■	■	■
Haryana					■	■					■	■
Himachal Pradesh						■						
Karnataka	■		■	■			■	■	■	■	■	
Madhya Pradesh	■	■	■	■						■	■	■
Maharashtra	■	■	■	■	■				■	■	■	■
Orissa	■		■	■								
Punjab					■	■					■	■
Rajasthan				■	■					■	■	■
Tamil Nadu			■	■			■	■	■	■	■	
Uttar Pradesh					■	■					■	■
Uttarakhand						■						
West Bengal		■	■	■								

■ Available season

Figure 1: Round the Year Availability of Onion in India

the harvest is due to loss in weight since onion contains 80-85 per cent moisture. The factors causing postharvest losses are mechanical injuries, improper drying/curing, sprouting, decay and weight loss (shriveling). Mechanical injuries are caused when too large bags are used, which are difficult to handle and cause mechanical damage to onion during loading and unloading for marketing. Drying/Curing of onion plays a vital role in reducing inpostharvest losses of onion. The purpose of drying is to remove excess moisture from the outer skin and neck region of the onions that prevents infection by disease causing organisms while minimizing shrinking caused by removal of excess moisture from the interiors. Sprouting of onion greatly reduces their utilization, value and accelerates deterioration. Harvesting of bulbs at premature stage increases changes of weight loss and sprouting during storage. Similarly, close neck cut of bulbs also aggravate postharvest decay. Various preharvest crop management practices *viz.*, irrigation, fertilizer and chemical application, time and stage of harvesting affect postharvest losses. Further, curing, sorting, grading, packing; storage and transportation are the major factors responsible for quality ensuring and reducing losses of the harvested onion.

4.1 Effect of Preharvest Factors on Postharvest Management

4.1.1 Varieties

Light red varieties *e.g.* Agrifound Light Red, ArkaNiketan, N-2-4-1, Pusa Red and NHRDF Red-2, generally grown during *rabi* season are good for storage. Dark red varieties in view of having lower TSS and dry matter content as well as less number of dried outer scales grown in *kharif* season have poor storability compared to light red varieties.

Table 3: The List of Onion Varieties Developed By Different Agencies in the Country

Name of Variety	*Developed by*	*Season*
	COMMON BIG ONION (*Allium cepa* L.) Red Coloured Bulbs	
Agrifound Dark Red	NHRDF, Nashik	*Kharif*
Agrifound Light Red	NHRDF, Nashik	*Late Kharif and Rabi*
Arka Kalyan	IIHR, Bangalore	*Kharif*
Arka Niketan	IIHR, Bangalore	*Late Kharif and Rabi*
Arka Pragati	IIHR, Bangalore	*Rabi*
Baswant -780	MPKV, Rahuri (MS)	*Kharif*
Bheema Raj	DOGR, Rajgurunagar	*Kharif*
Bheema Red	DOGR, Rajgurunagar	*Kharif*
Bheema Shakti	DOGR, Rajgurunagar	*Rabi*
Bheema Super	DOGR, Rajgurunagar	*Kharif and Late kharif*
Bheema Kiran	DOGR, Rajgurunagar	*Rabi*
Hissar-2	HAU, Hissar	*Rabi*
Kalyanpur Red Round	CSAUAT, Kanpur	*Rabi*
N-2-4-1	Dept. of Agril.(MS)	*Late Kharif and Rabi*
N-53	Dept. of Agril. (MS)	*Kharif*
NHRDF Red (L-28)	NHRDF, Nashik	*Rabi*
NHRDF Red-2 (L-355)	NHRDF, Nashik	*Rabi*
NHRDF Red-3 (L-652)	NHRDF, Nashik	*Rabi*
NHRDF Red-4 (L-744)	NHRDF, Nashik	Late *Kharif and Rabi*
Phule Samarth	MPKV, Rahuri (MS)	*Late Kharif*
Punjab Red Round	PAU, Ludhiana	*Rabi*
Punjab Selection	PAU, Ludhiana	*Rabi*
Pusa Red	IARI, New Delhi	*Late Kharif and Rabi*
Pusa Madhavi	IARI, New Delhi	*Rabi*
Pusa Ratnar	NBPGR, New Delhi	*Rabi*
Udaipur -101	RAU, Bikaner Campus, Udaipur	*Rabi*
Udaipur -103	RAU, Bikaner Campur, Udaipur	*Rabi*
VL – Piaz - 3	VPKAS, Almora	*Rabi*
	Yellow Coloured Bulbs	
Arka Pitamber	IIHR, Bangalore	*Rabi*
Early Grano	IARI, New Delhi	*Late Kharif and Rabi*
Phule Suvarna	MPKV, Rahuri (M.S)	*Late Kharif and Rabi*
Spanish Brown	IARI, RRS, Katrain, Kullu (HP)	In Hills

Contd...

Table 3–*Contd...*

Name of Variety	*Developed by*	*Season*
	White Coloured Bulbs	
Agrifound White	NHRDF, Nashik	*Rabi*
Akola Safed	PKV, Akola (M.S)	*Rabi*
Bheema Subra	DOGR, Rajgurunagar	*Rabi*
JNDWO	GAU, Junagadh	*Rabi*
N-257-9-1	Dept. of Agril.)M.S)	*Rabi*
Phule Safed	MPKV, Rahuri	*Rabi*
Punjab-48	PAU, Ludhiana	*Rabi*
Pusa White Flat	IARI, New Delhi	*Rabi*
Pusa White Round	IARI, New Delhi	*Rabi*
Udaipur-102	RAU, Bikaner Campur, Udaipur	*Rabi*
	SMALL COMMON ONION	
Agrifound Rose	NHRDF, Nashik	*Kharif in Andhra and All 3 seasons in Karnataka*
Arka Bindu	IIHR, Bangalore	*All 3 seasons in Karnataka*
Arka Swatista	IIHR, Bangalore	*White coloured – Kharif and Rabi In Karnataka*
	MULTIPLIER ONION (*Allium cepa* var. *aggregatum*)	
Agrifound Red	NHRDF, Nashik	*Kharif and Rabi*
Arka Ujjwal	IIHR, Bangalore	*Dark Red Multiplier type for Karnataka*
Co-1	TNAU, Coimbatore	*Kharif and Rabi*
Co-2	TNAU, Coimbatore	*Kharif and Rabi*
Co-3	TNAU, Coimbatore	*Kharif and Rabi*
Co-4	TNAU, Coimbatore	*Kharif and Rabi*
Co-On-5	TNAU, Coimbatore	*Kharif and Rabi (Mutlore type)*
MDU-1	TNAU, Coimbatore	*Rabi*
	ONION HYBRIDS	
Arka Kirtiman	IIHR, Bangalore	*Rabi*
Arka Lalime	IIHR, Bangalore	*Rabi*
Hybrid - 35	IARI, New Delhi	*Rabi*
Hybrid - 63	IARI, New Delhi	*Rabi*

4.1.2 Seed

Good quality seed from genuine sources should be used for getting good quality bulbs free from doubles, bolters, thick neck and having long shelf-life.

4.1.3 Soil

It has been observed that the onion grown in sandy soil does not produce compact bulbs with high TSS as compared to the crop grown in clay loam soil. Hence clay loam soil should be preferred for raising the crop meant for long storage or long distance transportation.

4.1.4 Fertilizer

FYM @ 25 tonnes per hectare and Nitrogen @ 100 kg per hectare should be applied for getting bulbs with good storable quality. This will, however, depend upon the nutrient status of soil. Potassium @ 50 kg per hectare is recommended for production of good quality storable bulbs. Top dressing with N should be completed before 60 days after transplanting so as to have thin necks. High dose of fertilizer (N: K: S) increases weight loss, rotting and sprouting percentage. Increase of N doses leads to rotting when stored in natural conditions. Use of calcium ammonium nitrate as N source has been reported to improve the bulb shelf life of bulbs.

Supplementary application of organic fertilizers comprising of 3 per cent panchagavya + 50 per cent FYM + 50 per cent poultry manure recorded better postharvest quality of onion.

4.1.5 Irrigation

Light and frequent irrigations are necessary. Heavy watering must be avoided at all costs. Irrigation at 10-15 days interval should normally be given. It should be stopped 10-15 days before harvesting. Uprooting of bulbs immediately after irrigation reduces their storage life. Irrigation after the crop has reached maturity results in re-sprouting and doubles which increases losses in storage. The irrigation through drip has been reported to be better compared to flood irrigation in reducing the storage losses.

5.0 Postharvest Management

5.1 Preharvest Treatments

Bavistin @ 0.1 per cent sprayed 90 days after transplanting or Klorocin @ 25 kg per hectare as preharvest treatment reduces spoilage of harvested produce.

5.2 Harvesting

The harvesting of onion bulbs should be done after one week of 50 per cent neck fall in *rabi* season. In *kharif* season when bulbs have fully developed and tops of the leaves have started yellowing, bulbs should be considered ready for harvesting. Bulbs harvested at premature stage does not undergo dormant stage and incur more losses due to weight loss and sprouting. Delay in harvesting results in splitting and bolting of bulbs. Such bulbs do not store well in storage.

No doubt production and productivity have increased substantially over the past 20 years; it is still quite low when compared to the countries like Netherlands, United States of America, China and Korean Republic. The constraints which have been identified for low production and productivity are enumerated as under:

i) Non-availability of high yielding storable varieties for all the three seasons.

ii) Lack of strategy for expansion of area in kharif, late kharif and rabi seasons in non-traditional pockets.

iii) Non-availability of quality seed of improved varieties in adequate quantities.

iv) Onion varieties developed are not uniform in colour, shape and size. Bolting and doubles are observed in almost all varieties. The marketable recovery thus is low.

v) Non-availability of high yielding and storable F1 hybrids.

vi) Susceptibility to diseases and insect pests.

vii) Sub-optimal standards of cultivation adopted by farmers.

viii) Non-availability of high TSS white onion varieties suitable for dehydration and high TSS bigger bulb sized yellow varieties for export.

ix) Inadequate storage facilities, transportation and marketing support.

5.3 Top Cutting

Cutting of tops close to bulbs cause entry of moisture and thereby more losses are caused due to decay. The NHRDF has standardized the technology and recommended to cut the tops of harvested onion at 2-2.5 cm above the bulbs so as to have closed neck.

5.4 Drying and Curing

The purpose of drying is to remove excess moisture from the outer skin and neck region. Curing is an additional process of drying to develop skin colour. Field curing for 3-5 days along with the tops by windrow method followed by 10-12 days of curing in shade before storage for *rabi* crop is necessary to minimize the losses. In *kharif* season if temperature at harvest is low, 10-15 days sun curing may be desirable but which is not always possible.

The curing of kharif onion is very difficult in rainy season due to cloudy conditions. The NHRDF has developed artificial curing chamber for curing of kharif onion. The temperature and wind velocity has been standardized and bulbs in such chamber with full load at 35 °C temperature and velocity of airflow at 3.2 m s^{-1} cured perfectly in 16 hours for marketing.

5.5 Sorting and Grading

The diseased, thick necked, damaged, bolted and double bulbs should be removed at the time of sorting and grading. The medium sized bulbs of 4-6 cm size are good for storage. Sorting and grading is mostly done manually by the traders for domestic as well as export supplies. However the NHRDF has developed with the help of a private party at Meerut electrically operated commercial graders, which are available for grading of onion in different sizes (Figure 1).

5.6 Packaging

Onions are often brought loose from farmer's field to market yard by bullock carts, tractor-trolleys. In some markets farmers bring onion in perforated 40 kg capacity jute bags of 9x10 mesh per sq inch and plastic woven bags having enough ventilation. Small pickling onion and multiplier onions are packed in 14-15 kg wooden baskets/bamboo baskets. For export, common big onions are packed in 5 to 25 kg size open mesh jute bags (Figure 2). However packaging been improved now and plastic leno bags are gradually replacing the traditional jute bags. In some countries plastic woven bags with good labeling and printed information about crop, variety, weight, the address of supplier and other details are commonly shared. Use of netted leno bags has also been started on a small scale for domestic market as also for temporary storage. Plastic woven bags are not only attractive but uniformity of onions when graded, could also be easily seen from outside.

Figure 1: Onion Grader

5.7 Storage

Every agricultural commodity is required to be stored properly to prolong the availability with minimum qualitative and quantitative losses. Onion is not an exception. The onion bulb is a natural food store for the plant, but it is a living system undergoing a process of development towards sprouting, and is subject to decay by various disease causing organisms. The objective of appropriate storage technology is to retain the quality of bulbs for as long as possible and for long distance marketing after removal from store, without much loss.

Figure 2: Packaging of Onion

The onion bulbs are generally stored from April/May to October/November for a period of four to six months. It has generally been noted that under normal conditions the total onion storage loss is around 15-40 per cent depending on the quality of stored material and duration of storage. The total storage losses comprise of physiological loss in weight (PLW) showing shrinkage (10-25 per cent), rotting (3-8 per cent) and sprouting (2-10 per cent). The PLW can be minimized by harvesting at right time, proper curing, and subsequent storage at desired temperature and humidity conditions. Generally, the rotting losses are at peak in initial months of storage, particularly in the months of June and July, due to prevalence of high temperature and high humidity. However, proper grading and selection of quality bulbs and good ventilation in storage condition can reduce rotting.

It is necessary to have the knowledge of the physiology of dormancy and epidemiology of storage diseases while deciding long term storage protocol. Systems thatprovide long dormancy and suitable condition unfavorable for disease development can be engineered using the physical principles of temperature and humidity control.

The physiological and pathological processes that proceed within a store of onion interact with the physical process of heat and water vapour exchange and mutually influence the environment within the store. Main factors, which influence onion storage and bring change in the bulbs, are summarized as under.

i) Sprouting and internal root development change bulb shape crack the skins and increase water loss.

ii) Sprouting increases respiratory rates of onion bulb that leads to.

iii) Generation of heat, CO_2.

iv) Bulbs thus get deteriorated due to disease development.

v) Bulb deterioration due to diseases increases respiratory outputs. The onion skins has vital role in the physical and physiological processes in the storage, as it is the main barrier to water loss and CO_2 exchange. A relative humidity of 65-70 per cent is desirable to maintain the skin fairly flexible and elastic. At lower RH, the skin become very brittle and gets easily cracked notably when skin moisture content falls below 20 per cent.

vi) Ventilation is another key element for maintaining humidity in storage. Lack of ventilation often adversely affects the quality by increasing bulb respiration and heat accumulation.

vii) With the advancement in storage period requirement of ventilation increases

viii) Design of onion store should, therefore, match these requirements.

ix) High humidity with high temperature favors spread of pathogens within the store.

It is necessary to counter the above changes by proper monitoring of internal environment. Heat and water vapor must be removed or introduced as necessary either by using heating or refrigeration or ventilation or a combination of all the mechanisms. However, under Indian conditions in onion growing states designs to exploit natural ventilation is most economical.

5.8 Transportation

Onions are transported in bullock carts, tractor-trolleys to local markets. Trucks and railway wagons are used for long distance movement within the country. Ventilated containers are mostly used for export to the neighboring countries. A few exporters now started using refrigerated container for export of onion to different countries especially to European countries.

6.0 Domestic Marketing and Export

Most of the farmers bring onion directly to the market after harvest, as proper storage facilities are not available with them. Fearing losses, farmers usually sell their entire stock within a month of harvest. As a result, during this period prices rule very low due to glut situation in the market. Thereafter, the rise in prices is quite rapid and sometimes-wide fluctuations occurs leading to dissatisfaction amongst the producers as well as consumers. There is a quantum jump in the export of onion during the last 3 decades (from 169771 tonnes during 1981-82 to 18.23 lakh tonnes during 2012-13) increasing the foreign exchange earnings from Rs. 29.43 to 2294.91crores.

Farmers generally sell their produce through regulated assembly/wholesale markets in major producing areas in different states. A number of agencies including retailers, wholesalers, commission agents and cooperatives are involved in marketing of onions. Cooperatives have been playing a significant role in the marketing of onion particularly after NAFED's entry in this trade. Cooperatives at district, talukas and market level act as commission agents for sale of the produce its members. Lasalgaon, Pimpalgaon, Chandwad, Manmad in Nasik, Rahuri in Ahmed Nagar, Lonand in Satara and Chakan in Pune district in Maharashtra are the major assembly markets in producing centers. Mahuva, Bhavnagar and Gondal in Gujarat, Bangalore, Hubli, Belgaon in Karnataka, Kurnool in A.P. are the major assembly markets in major producing pockets in respective states. From Rajasthan and Haryana onions are directly brought to Delhi or other places for local distribution.

Continued export fetches good returns to farmers. Open auction system, under cover, tender system and open agreement of sales are followed in different assembly and distributing markets of the onions in the country. The farmers send their produce to commission agents who arrange sale in the markets. After purchase the traders and exporters take up sorting, grading and packing and dispatch to different destinations according to the demand.

It has been observed that in many areas regulated system of marketing does not exist. Informal group of farmers or cooperatives are being formed at village level for helping in marketing of produce at reasonable price.

6.1 Production for Export

Continuity in onion export is must to maintain the present status of production in the country and maintain the domestic prices at remunerative level. Regularity in export also maintains the confidence of foreign buyers towards Indian onions.

Presently 90 per cent of export is comprised of mainly common big onion types grown in Maharashtra near Nasik, Pune, Ahmednagar, Satara etc.; Dindigul and Coimbatore in Tamil Nadu and Mahuva (Bhavnagar) in Gujarat and Patna and Biharsharif in Bihar. Small pickling onions grown near Cuddapah in A.P. near Bangalore and Chickballapur in Karnataka and multiplier onion gown near Dindigul are also exported. Agrifound Dark Red, N-53, local Bellary Red, Patna Red, Agrifound Light Red and local Poona Red falls in this group. The big onions are exported to Gulf countries, Sri Lanka, Bangladesh, Singapore and Malaysia. Small onions such as Agrifound Rose and multiplier onions like CO4, CO3, COOn5 and Agrifound Red contributes 10 per cent of Indian export. Small onions are exported to Bangladesh. Medium sized are exported to Singapore and Malaysia.

Yellow onion, though are in demand in European and Japanese markets, and are not being grown on commercial scale in India. Some yellow onion varieties have been tested by NHRDF and can be grown successfully for export purpose. There is wide scope for yellow onion export to European countries. It is however, necessary to ship in refrigerated or ventilated containers since these short day yellow onions have poor keeping quality. The NHRDF has tested yellow onion varieties *i.e.* Tana F1, Suprex, Granex-33, Granex-429, Mercedes, Excalibur, Equinex, Lexus, Cougar, Linda Vista, and Colina at its Regional Research Stations Nashik, Talaja(Bhavnagar) and Karnal. The varieties Excalibur, Mercedes, Cougur and Colina performed better by recording more than 70 per cent recovery of 6.50 cm and above size of bulbs which is generally preferred in European countries.

NHRDF had also tested many white onion hybrid varieties *i.e.* Rio Jefe(H), White Creole, Perla, Estrella Blanco, Early Lockyer White, SSC-6008(H), Kristel, at its Regional Research Stations Nasik, Talaja and Karnal. The variety White Creole and Kristel performed better. White Creole recorded high TSS 16-18 per cent, while Kristel recorded TSS 9-11 per cent. Onion variety Kristal is suitable for fresh export to European countries and commonly grown in Gujarat under contract farming. The agro-techniques of above recommended varieties have also been developed by NHRDF.

White and yellow onions are not preferred for consumption in many areas in India, due to less pungency. These are mainly used exclusively for export or by dehydration industries. Therefore production of high TSS white onion varieties and yellow/red onion varieties could be increased by contract farming in Maharashtra and Gujarat. There is need to increase the big onion production in Maharashtra and Gujarat and pickling type and small onion production in Andhra and Karnataka and multiplier onion production in Tamilnadu to increase the onion export from India.

6.2 Packaging for Exports

Packing size wholly depends on the requirement of importing country. Some of the netted poly bag sizes have been given below for different countries. Packaging and size requirement for export of onion to various countries are mentioned in Tables 5 and 6.

Table 5: Packaging Requirement of Export of Onion to different Countries

Importing Nation	*Pack Size*
Malaysia and Indonesia	10 kg
Srilanka	25 kg
Singapore	8 kg
Philippines	20 kg
Gulf countries	5 kg, 10 kg, 15 kg and 20 kg

Source: NIAM, Jaipur.

Table 6: Size of Onion Required by different Countries

Country	*Size (mm)*
Middle East and Gulf (Dubai, Sharjah, Doha, Muscat, Bahrain, Dammam, Saudi Arabia, Kuwait)	40 – 60
Malaysia, Singapore, Port Kelang and African Ports	25 – 30
Sri Lanka	25 – 30
Bangladesh, Pakistan and Nepal	25 – 30
Europe	60 – 70
Japan	60 – 70

Source: MSAMB, Pune.

6.3 Export Marketing

Global trade of onion export is 7.14 million tons during 2010. Netherland is the leading exporter (1.50 million tons) of onion contributing to about 21 per cent of world export. India, China, Egypt and USA are at 2nd, 3rd, 4th and 5th position contributing to about 19 per cent, 9 per cent, 6 per cent and 5 per cent share respectively in world export (FAO 2010). However, during the year 2009 India was the major exporting country in the world. Russian Federation, Bangladesh, Malaysia, USA, UK, Japan, Saudi Arabia, Germany, UAE, Netherlands, Canada and Pakistan are major onion importing countries in the world.

Even in 1951-52, the country exported more than 5000 tonnes of onion per annum. In 60 (s) the export was almost doubled and further in 80(s) it jumped manifold touching 2.74 lakh tonnes. Onion exports touched new height during 2009-10 by exporting 18.73 lakh tonnes of value, Rs.2834 crores. However, the export to some extent came down to 13.41 lakhs tones with value of 2159.06 crores during 2010-11 due to less productionof *kharif* onion and ban on export.The onion export again affected during the year 2011-12 and reached only upto 15.52 lakhs tones with value of Rs. 2141.43 crores due to less production of *kharif* onion and ban on onion export. Presently onion is exported to Bangladesh, Malaysia, Dubai, Qatar, Kuwait, Singapore, Sri Lanka, Mauritius and Nepal. Onion export in earlier days was mediated through National Agricultural Cooperative Marketing Federation of India (NAFED). Carrying 13 state agencies including NAFED are involed in facilitating the export of onion from India.

However NAFED is still the nodal agency, responsible for monitoring the export through other canalizing agencies. NAFED also exports onion on its own. The export is mainly from Maharashtra and Tamil Nadu ports. Export is also made by road to Bangladesh and Nepal from UP, Bihar and West Bengal.

6.4 Grade Designation and Quality of Onions for Export

Government of India has prescribed certain grades for exportable onions according to which the Onion should be:

i) Intact,

ii) Sound produce affected by rotting or deterioration which make it unfit for consumption is excluded.

iii) Clean, free from any visible foreign matter,

iv) Free from damage caused by frost;

v) Sufficiently dry for the intended use (in the case of onions for storing, at least the two first outer skins and the stem must be fully dried).

vi) Without hollow or tough stems;

vii) Practically free from damage caused by pests;

viii) Free of all abnormal external moisture;

ix) Free of any foreign smell and/or taste.

x) The stem must be twisted or clean cut and must not exceed 6 cm in length (except for stringed onions).

xi) They shall comply with the residue level of heavy metals, pesticides and other food safety parameters as laid down by the Codex Alimentarius Commission for exports.

As per AGMARK onion is classified under three classes *viz.* Extra class, Class-I and Class-II, the details of which are mentioned as under (Table 7).

Table 7: AGMARK Grades and Standards for Onion

Grade Designation	*Grade Requirements*	*Grade Tolerance*
Extra Class	Onion shall be of superior quality. They shall be characteristic of the variety and/or commercial type. The bulbs shall be: ☆ Firm and compact; ☆ Unsprouted (free from externally visible shoots); ☆ Properly cleaned; ☆ Free from swelling caused by abnormal development; ☆ Free of root tufts, however, in onions harvested before complete maturity, root tufts are allowed. They shall be free of defects, with the exception of very slight superficial blemishes, provided these do not affect the general appearance of the produce, the quality, the keeping quality and presentation in the package.	5 per cent by number or weight of onion not satisfying the requirements of the grade, but meeting those of class I grade or, exceptionally, coming within the tolerances of that grade.
Class – I	*Onion shall be of good quality and characteristics of the variety and/or commercial type. The bulbs shall be: ☆ Firm and compact; ☆ Unsprouted (free from externally visible shoots); ☆ Properly cleaned; ☆ Free from swelling caused by abnormal development; ☆ Free of root tufts, however, in onions harvested before complete maturity, root tufts are allowed.	10 per cent by number or weight of onion not satisfying the requirements of the grade, but meeting those of class II grade or, exceptionally, coming within the tolerances of that grade.

Contd...

Table 7–*Contd...*

Grade Designation	Grade Requirements	Grade Tolerance
	The following slight defects, however, may be allowed, provided these do not affect the general appearance of the produce, the quality, the keeping quality and presentation in the package. ☆ A slight defect in shape. ☆ A slight defect in colouring; ☆ Light staining which does not affect the last dried skin protecting the flesh, provided it does not cover more than one-fifth of the bulb's surface. ☆ Superficial cracks and absence of part of the outer skins, provided the flesh is protected.	
Class – II	Onions which do not qualify for inclusion in higher grade but satisfy the minimum requirements. They shall be reasonably firm. The following defects, however, may be allowed, provided the onion retain their essential characteristics as regard the quality, the keeping quality and presentation. ☆ Defects in shape; ☆ Defects in colouring; ☆ Early signs of shoot growth visible from outside(not more than 10% by number or weight per unit of presentation); ☆ Traces of rubbing; ☆ Slight marking caused by parasites or disease; ☆ Small healed cracks; ☆ Slight bruising, healed, unlikely to impair keeping qualities; ☆ Root tufts; ☆ Stains which do not affect the last dried skin protecting the flesh provided they do not cover more than half the bulb's surface; ☆ Cracks in the outer skins and the absence of a part of the outer skins from not more than one third of the bulb's surface, provided the flesh is not damaged.	10 per cent by number or weight of onion not satisfying the requirements of the grade but meeting the minimum requirements.

a) Size Grading as per AGMARK

Size is determined by the maximum diameter of the equatorial section in accordance with the Table 8.

Table 8: Size Grading of Onion

Size Code	Diameter (mm)	Difference between the Diameter of the Smallest and Largest Onion in the Same Package (mm)
A	10—20	5
B	21—40	15
C	41-70	20
D	71 and above	30

Size tolerances: For all grades: 10 per cent by weight of onions not satisfying the size identified, but with a diameter of no more than 20 per cent below or above it.

7.0 Conclusion

Onion is one of the most important commodities that play critical role in the agri-based economy of many parts of India due to its great demand in India and abroad. However, awareness on postharvest management of this crop among the growers, wholesalers and retailers is very important for saving the huge losses after harvest. An integrated approach for reduction in postharvest losses is important. This should begin with development and promotion of varieties, refinement in agro-techniques with respect to production and crop protection and adoption of appropriate postharvest handling technologies.

The postharvest life of onion produced in any season largely depends upon curing of the bulbs, which involves field curing as well as shade curing. These two operations can be very well performed in late *kharif* and *rabi* crops, however, the kharif crop harvested in the month of October-November hardly gets an opportunity for proper curing due to extended monsoon in western zones and due to low temperature in north zones. On-farm artificial curing facilities can resolve these problems. Similarly, large size godowns with forced ventilation and multi-tier system will be the future requirement. Standardization and designing godowns with 1000 tonnes capacity for traders and exporters are also the need of the hour.The marketable onions both for export and domestic markets suffer from poor grading. The real benefits of market price are masked due to improper/no grading and unattractive packing. The concept of mechanized grading is totally missing in the onion trade in the absence of proper graders at affordable prices. There is need for development of on-farm motorized graders with a capacity of 2-5 tonnes/hr.

8.0 Important References

Bhonde, S.R., Srivastava, K.J., Sharma, H.K and Chougule, A.B., 1996. Effect of withholding irrigation before harvesting on storage life of onion. *NHRDF News Letter*,**16 (4)**: 1-14.

Bhonde,S.R., Ram, L. and Pandey,U.B. 1983. Effect of curing methods and stage of harvesting on storage quality of onion. **In:** *Proceedings of National Workshop on onion* organised by NAFEDin collaboration with Min. of Agriculture, Govt. of India,Dec., 17-18th pp 13-22.

Chauhan, K.P.S., and S.P. Singh and Chougule, A.B. 1995. Studies on the effect of windrow curing, neck cutting and shade curing on export quality of onion bulbs during storage. *NHRDF News Letter*,**15(4)**: 5-7.

Cho Jungeun, Bare Rona, Lee Seungkoo, 2010. Current research status of postharvest technology of onion (*Allium cepa* L.) *Korean J. Hort.Sci. and Tech.*, **28 (3):** 522-527.

Chougule, R.R., Bhonde, S.R. and Pandey, U.B. 2004. Assessment of postharvest losses in onion. *NHRDF News Letter*, **24(1):** 11-16.

Deb, M., Pandit, M.K., Sarkar, A. and Maity, T.K. 2009. Response of irrigation and fertilizer on growth, yield and postharvest shelf-life of onion. *Env. and Eco.*,**27 (3):** 1025-1027 Calcutta, India.

Gupta. R.P. and Srivastava, P.K. 1984. A study of storage diseases of onion. *Indian Phytopath.*, **37 (2):** 404.

Sankar, V.,Veeraragavathatham, D. and Kannan, M. 2009. Postharvest storage life of onion influenced by organic farming practices farming practices. *Asian J.Hort.*,**4(1):** 16-20.

Singh, D.K., Pandey, A.K., Pandey, U.B. and Bhonde, S.R. 2002. Effect of farm yard manure combined with foliar application of NPK mixture and micro nutrient on yield, quality and storage of onion variety Agrifound Light Red. **In:** *Role of Technology in improving the productivity of small farmer Agriculture-A8* NIRD Foundation Day Seminar on Rural Technology for poverty alleviation, Hyderabad, January 2-3, concept paper and Abstract pp. 28.

Zeka, N., Nasto, T.H. and Balliu, A. 2008. The influence of N sources and fertilization doses on dry onion yield and bulbs shelf life. **In**: *proceedings of the IV Balkan Symposium on vegetables and potatoes*, Plovdiv, Bulgaria, 9-12 September 2008.

Chapter 34
Orchids

L.C. De

Principal Scientist, ICAR-National Research Centre for Orchids, Pakyong – 737 106, Sikkim, India

N. Pathak

Sikkim Himalayan Orchids Ltd., Assam Lingzey – 737 135, East Sikkim, India

E-mail: lcde_horti@rediffmail.com; E-mail: narender.pathak@gmail.com

1.0 Introduction

Orchids comprise the largest family of flowering plants with 25,000 to 35000 species belonging to 600-800 genera. They are prized for their incredible diversity in size, shape, form and colour, attractiveness of their flowers and high keeping qualities even upto 10 weeks. Most of the orchids have originated from tropical humid forests of Central and South America, India, Sri Lanka, Burma, South China, Thailand, Malayasia, Philippines, New Guinea and Australia. Brazilian *Cattleya*, Mexican *Laelia* and Indian *Cymbidium*, *Vanda* and *Dendrobium* have played a major role in developing present day beautiful hybrid orchids which number more than 2,00,000. In the international trade, among top ten cut flowers, orchids rank the sixth position and it occupies 3 per cent of the total cut flower production. *Cymbidium* holds the first rank among the orchids.

2.0 Status of Cultivation

In India, orchids comprising of 158 genera and 1331 species grow upto an elevation of 5000m. Indian terrestrials are naturally distributed in humus rich moist earth under tree shades in North Western India. Western Ghats harbour the small flowered orchids. Epiphytic orchids are common in North eastern India which grow upto an elevation of 2000m from sea level. Indian orchid species having high ornamental values are *Aerides multiflorum, Aerides odoratum, Arundina graminifolia, Arachnis, Bulbophyllum, Calanthe masuca, Coelogyne elata, Coelogyne flavida, C. corymbosa; Cymbidium aloifolium, C. lowianum, C. devonianum, C. hookerianum, C. lancifolium; Dendrobium aphyllum, D. nobile, D. chrysanthum, D. farmeri, D. chrysanthum, D. densiflorum, D. moschatum, D. fimbriatum, D. jenkinsii; Paphiopedilum venustum, P. spicerianum, P. hirsutissimum, P. insigne, Phaius wallichii, Pleione praecox, Renanthera imschootiana, Rhyncostylis retusa, Thunia alba, Vanda cristata, Vanda coerulea* and *Vanda coerulescens*. Varied agroclimatic conditions of our country is highly favourable for cultivation of commercial orchids like *Cattleyas, Vanda, Dendrobium, Arachnis* and *Aerides* in tropical areas, *Cymbidium, Paphiopedilum, Dendrobium* and *Phalaenopsis* hybrids in subtropical areas and *Cypripediums, Cymbidiums* etc in temperate areas. Among orchids, *Cymbidium, Dendrobium, Phalaenopsis, Cattleya, Vanda, Aranda, Arachnis* etc used as cut flowers whereas *Aerides, Arachnanthe, Bulbophyllum, Calanthe, Coelogyne, Eria, Phaius, Phalaenopsis, Pleione, Rhyncostylis* are used as potted orchids. The *Cymbidium* is mainly grown in NEH Region, Sikkim, Darjeeling hills and Arunachal Pradesh. Tropical orchids are cultivated in Kerala and some parts of Tamil Nadu. We should prefer those species which flower during winter and spring months to export flowers to temperate regions from December to May.

India accounts for nearly 7 per cent of worlds orchid biodiversity which is distributed in five major phyto-geographical regions *viz.* North eastern Himalayas, Peninsular region, Western Himalayas, Western Ghat and Andaman and Nicobar Islands (Table 1).

Table 1: Status of Orchid Distribution in India

Region	*States*	*Genera*	*Species*
Andaman & Nicobar Islands	319 Islands and islets in the Bay of Bengal	53	115
Eastern Himalayas and North Eastern India	Arunachal Pradesh, Assam, Manipur, Meghalaya, Mizoram, Nagaland, Sikkim and Tripura	159	870
North -Western Himalayas	J&K, Himachal Pradesh, Uttarakhand	75	288
Peninsular India	Andhra Pradesh, Central India, Eastern Ghats, Gujarat, M.P., Orissa, and Western Ghats	89	379

Source: Hajra and De, 2010.

3.0 Commercial Varieties

The commercial varieties of orchids grown belong to the following types.

a) Cymbidium

Cymbidiums are among the most popular winter and spring blooming semi-terrestrial orchids originated from tropical and subtropical Asia covering North

Eastern India, China, Japan, Malayasia, the Philippines, Borneo islands and North Australia, usually grown in cooler climates at high elevations. Cymbidiums are highly valued for cut flowers, hanging baskets, potted plants and herbal medicines. Cymbidium has been considered as top commercial orchids in Europe since many years. They fetch the highest price in the international markets of which major Asian markets of Singapore and Japan or the Dutch market.In India, Arunachal hills, Sikkim and Darjeeling hills with cool summer night and monsoon summer rains are ideal for cymbidium cultivation. In Sikkim, more than 350 hybrids of cymbidium orchids are commercially cultivated in an around 28 ha of land and about15-20 million spikes are produced annually.

b) Dendrobium

Dendrobiums are popular flowering potted plants and cut flowers around the world due to their floriferous nature, wide range in flower colour, size and shape, year round availability and long vase life. Hawaii, California and Florida are major potted *Dendrobium* growing regions in the United States. In the Netherlands, production of potted orchids is now 40 to 50 million units with *Dendrobium* increasing in popularity. Imports from Thailand, the world's largest exporter of tropical cut orchids and second largest supplier to the EU, accounted for 22 per cent of supplies to the EU. Thailand holds a particularly strong position in *Dendrobium* orchids.

c) Phalaenopsis

Phalaenopsis is the second most valuable and popular flowering potted plant and cut flower around the world due to their easy cultural practices, diversity in flower colour, size and shape, year round availability, delicacy and long vase life. It is commercially grown in Germany, Japan, The Netherlands, Taiwan and United States. In the United states, 75 per cent of all orchids purchased are *phalaenopsis.*

3.1 Other Tropical Orchids

Vanda is widely distributed throughout Australasia from China through the Philippines, Indonesia, Malaysia, New Guinea and Australia, Myanmar, Thailand, India and Sri Lanka. In the world tropical orchid trade, *Dendrobium* is the most dominant crop in addition to *Mokara, Oncidium, Aranthera, Aranda, Vanda, Arachnis, Renanthera, Ascocenda, Phalaenopsis, Cattleya* and *Paphiopedilum* which are being grown as cut flowers and potted plants.

3.2 Orchid Hybrids

Among the cut flower crops, *Cymbidium, Dendrobium, Phalaenopsis, Odontoglossum, Oncidium, Cattleya, Paphiopedilum, Vanda, Aeridovanda, Aranda, Mokara, Arachnis, Vascostylis, Renanthera, Rhyncicentrum, Rhyncovanda* etc. are important.

A good quality cut flower of an orchid should have the following characteristics.

i) Minimum eight standard blooms per stem

ii) Flowers must be cleaned, evenly coloured and free from physiological disorders

iii) Stem must have flowers evenly arranged and around the stem.

iv) Two third of the stem should be covered with the flowers.

v) Flowers must have a firm texture and a luminescent sheen

vi) Stems must be firm when held up

vii) The minimum base diameter of the stem should be of 10mm

Important varieties and hybrids under different genera of orchids for cut flower purposes are listed in Table 2.

Table 2: Important Hybrids under different Genera of Orchids

Genera	*Hybrids*
Aeridovanda	Doctor Poyck, Vieng Ping, 'Bensiri', 'Noreen', 'Early Bird', 'Shiv Sidhu', 'New Dawn', 'Harrison Luke Somsri Sunlight'
Aranda	'City of Singapore', 'Hilda Galistan', 'Urmila Nandey', 'Christine', 'Thailand Sunspot', 'Millenium Dawn', 'Broga Giant', 'Salaya Red', 'Propine White', 'Propin Spot', 'Lueng Cholburi', 'Ishbel Manisaki', 'Baytown', 'Chao Praya Blue', Chao Praya Dot Com', 'Chao Praya Beauty', ' Ethan Pride', 'Taksari Chandrabir', 'Thailand Sunspot'
Arachnis	'Ishabel', 'Maggie Oei', 'Maroon Maggie', 'Bartha Braga'
Ascocenda	'Apinantat Red Berry', 'Pralor Tuyen', 'Pak-Kred', 'Bangkok', 'Surin', 'Karnada', 'Crownfox', 'Sundancer', Laksi 'Red Ruby',Guo Chia Long 'Spotty', 'Fuchs Angel Frost', 'Carol Belk', 'Renuka Angle', 'Joyce Bevins', 'Adisak Blue', 'Renu Gold', 'Tipi Blue Boy', 'Bobs Fortune', 'Rubychai', 'Shah Rukh Khan', 'Yang Sophia Firuz', 'Abdul Ghani Othman', 'Chunika', 'Fuch's Star'
Cattleya and allied	'Lovely Bangkok', 'Admiration', 'Bob Belts', 'General Patton', 'Joyce Hannington', 'Little Angel', 'Margaret Stewart', 'Nillie Roberts', 'Pearl Harbour', 'Primma Donna', 'Queen Sirkhit' 'Diamond Crown', 'Secret Love', Ladda Belle 'Pink Pearl', 'Maikai', 'Pastoral', 'Robert', Prism Palette 'Tricolour Magic', 'Ahmad Seikhi', 'Hsinging Catherine', 'Chia Lin New City', 'Chinese Beauty Orchid Queen'
Cymbidium	'Levis Duke Bella Vista', 'Madrid Forest King', 'Sparkle Late Green', 'Angelica December Gold', 'Sleeping Nymph', 'Pine Clash Moon Venus', 'Soul Hunt', 'Dr. H. C. Aurora', 'Susan Highes', 'Tia Gaig Suther Land', 'Miss Sanders', 'Amesbury', 'Kenny Wine', 'Red Star', 'Red Princess', 'Show Girl', Jungfrau 'Snow Queen', Jungfrau 'Dos Pueblos', Lilian Stewart 'Coronation', Lilian Stewart 'Party Dress', Orkney 'Pink Heather', Ensikhan 'Alpha Orient', 'Fire Storm Ruby', 'Fire Storm Blaze', Bob Marlin 'Lucky'
Dendrobium	'Emma White', 'Thongchai Gold', Julie', 'Erika, Sonia-17', 'Sonia-28', 'Kasem White', 'Madam Pompadour', 'Bangkok Blue', 'Ann', 'Gold Twist', 'Candy Stripe Pink', 'Genting Blue', 'Bengal Beauty', 'Sakura Pink', 'Burana Charming', 'Blue Fairy', 'Channel', 'Nette White', 'Dang Saard', 'Big White 4N', 'Ear Sakul', 'Erika', 'Lervia'
Mokara	Walter Oumae 'Seksan', Thailand, Sayan, Walter Oumae 'Royal Sapphire', Susan 'Orange', Walter Oumae 'Calypso', Eng Ling, Madame Panne, Mak Chin On, Bangok Gold, Bibi, Chao Praya Gold, Chark Kuan Orange, Chark Kuan Pink, Chark Kuan Rose, Chark Kuan Super, Dinah Shore, Kelvin Red, Kelvin Orange, Luenberger Gold, Margaret Thatcher, Pink Star, Sayan, Sayang Pink, Walter Oumae, WTO, Jiti, Happy Beauty, Salaya Gold

Contd...

Table 2–*Contd...*

Genera	*Hybrids*
Odontoglossum	Carroll, Ismene, Palnina, 'Italian Job', 'Joyce Stewart', 'Pepe Gerald', 'Purple Rain', 'Katherine Jenkins', 'Roy Wittwer', 'Laguna Blanca', 'Precocious', 'Snow Fall', 'Pesky', 'Queen of Mars', 'Stam Point', 'Holiday Yellow', 'Ronald Norman', 'Bridget Ring Lawless', 'John Hanry Hanson', 'Point Pesky'
Oncidium	Aloha Iwanga Dogasima, Goldiana, Gower Ramsey, Golden Shower, Sum Lai Who Jungle Queen, Taka H and R, Sharry Baby Sweet Fragrance AM/AOS, Golden Glow, Popki Red, Irine Gleason Red, Vision Brownish Red, Catherine Wilson x New Calidonia Brownish Red, Robson Orchid Glad
Paphiopedilum	Niveum, Concolour, *P. rothschildianum* (3 to 5 flowers), *P. sanderianum* (3 to 5 flowers), Prince Edward of York, Michel Koopwitz, Saint Swithin, Mount Toro, Sorcerers Apprentice, Grande, Don Wimber, Elizabeth March, Hanne Popow, Jason Fischer, Living Fire
Phalaenopsis	Taisuco Crane, Taisuco Kochdian, Cygnus, Yukimai, Sogo Musadian, White Dream, Florida Snow, Nobby's Pink Lady, Minho Valentine, Minho King Beauty, New Cinderella, Taisuco Firebird, Sogo Smith, Carol Campbell, Emil Giles, Brother Lawrence, Taipei Gold, Golden Bells, Sogo Managers, Brother Passat, Be Glad, Cassandra, Vilind, Carmelas Pixie, Zuma's Pixie, Timothy Christopher, Be Tris, Quevedo, Strawberry, Detroit, Maki Watnabe, Kaleiodoscope
Renanthera	Brookle Chandler, Manila T-Orchids, Kilauea, Mok Yark-Seng, Poipu, Tom Thumb, 'Red Leopard', 'Scarlet Belle', 'Chanachae', 'Serdang', 'Brady Crocker', '20th WOC Singapore-2011', 'Bart Motes'
Renantanda	'Forever Yvonne', 'Inspiration Ng Teng Fong', 'Ladda Glow'; 'Polyetheramine Singapore', 'Momon Shija', 'Paul Gripp', 'Science Arts', 'Memoria Charles Darwin', 'Prof. G.J. Sharma', 'Kebisana Shija', 'Mary Motes', 'Kofi Annan'
Rhynchovanda	'Wilton Hill', 'Jammie Harper', 'Apichart', 'Noo Noi', 'Peter Draper', ' Brighton's Albino', 'Prairie Lady'
Vanda	Annette Jones, Antonio Real, Golamcos Blue Magic, Fuch's Charmer, Jimmy Millers RF Orchids, Keree Delight, Memoria Lyle Swanson, Motes Indigo x Merrillii, Motes Honeybun, Motes Primerose, Miss Joaquim,V. Rothschidiana, VTMA –Red, Pink, White, Vasco, Johnny Miller,Veerawan, Roberts Delight, Rasriprai, Pat Delight, Pakchong Blue, Mimi Plammer, Manuvade, Lumpini Red, Kultana Gold x Thongchai Gold, Fuchs Delight, Charles Goodfellow, Pine River,Adisak, Doctor Anek, John Club, Bill Sutton, Ellen Noa, Emily Notley, Evening Glow, Honomu, Honolulu, Hilo Blue
Vascostylis	Paragon Joy x Kasems Delight, Precious, Veeraphool, Crown Fox 'Red Yen', Aroon Fairy, Viboon Velvet, Chao Praya Lime', 'Lanna Rosy', 'Jeans Delight', 'Bay Sapphire', 'Spring Hill'

Source: Bhattacharjee and De, 2005; De *et al.*, 2013

4.0 Nature and Causes of Postharvest Losses

Opening of flower buds and rate of senescence accelerate at higher temperatures. At lower temperature, the respiration comes down and the flowers produce less amount of ethylene. Temperature plays an important role for flowers harvested at the immature stage for full expansion of buds and the flower buds are kept at temperatures as low as 0.5 to 4.0 °C in *Cymbidium* and *Paphiopedilum*, 5-7 °C in *Dendrobium* and 7-10 °C in *Cattleya*.

Light is essential for long distance transport or prolonged storage of cut flowers. Similarly, high light intensity is essential for opening of tight bud cut flowers. Florists had to maintain a light intensity of 2000-3000 lux for 12-24 hours in their shops for illuminations for most of cut flowers.

Cut flowers are kept at 90-95 per cent relative humidity for maintaining turgidity. Flowers start showing wilting symptoms when they have lost 10-15 per cent of their fresh weight. The rate of transpiration from leaves is reduced with the increase of high relative humidity (Figure 1).

***Cymbidium*, 'Fire Storm Ruby'**

***Dendrobium*, 'Dang Saard'**

***Oncidium*, 'Sharry Baby Sweet Fragrance'**

***Vanda*, 'Roberts Delight Blue'**

***Phalaenopsis*, 'Strawberry'**

Figure 1: Commercial Orchids Varieties

5.0 Postharvest Management

Postharvest life of orchid cut flowers is influenced by preharvest factor like varietal differences, light intensity, sugar level of flowers, temperature, nutrition and water loss; harvest factors like time, method and stage of harvest and postharvest factors including ethylene production, precooling, water quality, pulsing, use of preservatives, bud opening, packaging and storage. It is estimated that a third of the postharvest life of flowers is programmed by preharvest handling. Varietal differences in cut flowers had been reported due to variations in water uptake, fresh weight, flower diameter, stem lignification, vase life and senescence behaviour. The vase life of cut flower species and hybrids ranges from 10 to 56 days (Table 3).

Table 3: Vase Life of Orchids for different Species and Varieties

Name of Species/Hybrids	*Vase Life (Days)*
Aerides odoratum, Aerides multiflorum, Cymbidium iridioides, Dendrobium nobile, Renanthera imschootiana	28-56 days
Aranda	18-28 days
Cattleya hybrids	10- 20 days
Cymbidium hybrids	20-55 days
Dendrobium hybrids	14-21 days
Paphiopedilum hirtussimum, P. wardianum	56 days
Phalaenopsis hybrids	25-30 days
Phaius tankervillae	28-42 days
Vanda coerulea, Vanda teres, Zygopetalum intermedium	14-21 days
Vanda, Mokara hybrids	14-30 days

Source: Bhattacharjee and De, 2005

5.1 Harvesting

Flowers should be harvested in mild temperature because high temperature causes rapid respiration rates and excessive water loss. Flowers should be harvested in the early morning or in the evening. In the early morning, flowers remain turgid due to transpiration at night and higher sugar levels. Similarly, flowering stems retain a higher amount of stored carbohydrates if cut in the afternoon and retain more vase life.

Sharp tools or secateurs are always used to detach the stem of flowers from the mother plant. The angle of the cut is given in slanting position and the stem isnot crushed during harvesting, especially hard wood stems. The spikes aredipped in a bucket containing water immediately after harvest.

The optimum harvesting stage of the commercial orchids is fully open and mature flowers. The stage of harvest, spike length and number of flowers of some commercial orchids are given in Table 4.

Table 4: Stage of Harvest, Spike Length and No. of Flowers of some Commercial Orchids

Orchid Hybrid	Stage of Harvest	Spike Length (cm)	Flowers (No.)
Aranda	50 per cent bloom	45-60cm	8-10
Cattleya	2-4 days before bud open	25-40 cm	1 or more
Cymbidium	75 per cent bloom or two buds open stage	60-90 cm	10-15
Dendrobium	All flowers except top bud	40-60cm	8-12
Oncidium	80 per cent bloom	60 cm	Many
Paphiopedilum	3 to 4 days after opening of flowers	25-40 cm	1-5
Phalaenopsis	Fully open flowers	40-60 cm	8-10
Vanda	Fully open flowers	50-75 cm	8-15

5.2 Temperature

Generally, higher temperature results in higher level of respiration. Cooling is essential to reduce other metabolic changes such as enzymatic activity and to slow the maturation of flowers. Cooling prior to packaging and transport reduces ethylene production and improved longevity.

5.3 Light Intensity

Light is found to determine the carbohydrate levels before harvest which in turn influence the keeping quality. Flowers containing relatively higher amounts of carbohydrates especially mobile sugars lasted longer in the vase. Plants having few leaves, or leatherlike leaves (like most cattleyas and oncidiums), require a high-light environment. If the leaves are soft and limp (like some phalaenopsis and most paphiopedilum), the plants are probably very light-sensitive, and should not be placed in a sunny south-facing window. Most orchids prefer indirect or filtered light and 50 per cent shading. Some examples of light intensity requirements of orchds types are mentioned below:

i) Low light orchids (1200-2000 f.c.): *Phalaenopsis, Calanthe*

ii) Medium light orchids (2000-3000 f.c.): *Cattleya, Laelia, Brassovola*

iii) High light orchids (3000 f. c. or more): *Cymbidium*, Vandaceous groups

5.4 Precooling

It isan important operation in postharvest handling and transport of cut flowers for the fast removal of field heat wherever flowers are held dry pack. All flowers are precooled immediately after harvest by placing them in cold storage without packing or in open boxes until they reach the desired temperature. The temperature varies with the species and cultivars *e.g. Cattleya* (7-10 °C), *Cymbidium* and *Paphiopedilum* (0.5 to 4 °C), *Dendrobium* (5-7 °C). Precooling lowers respiration rate and decrease the break down of nutritional and other stored materials in the stems, leaves and petals; delays bud opening and flower senescence. It also prevents rapid water loss and decrease flower sensitivity to ethylene. Several precooling techniques such as room cooling, forced air cooling, hydro-cooling, vacuum cooling and ice bar cooling etc.

are available.The resistance to water flow through stem segments increase predominantly in the lower most one to two centimeter of the cut stems. Re-cutting stems under water improves longevity due to elimination of air from the conducting vessels.

5.5 Conditioning

Conditioning or hardening is a simple process where flowers are kept standing loosely in a big container so that air can circulate around the stems. The purpose of the treatment is to restore the turgidity of cut flowers with water stress during storage and transport. Conditioning is achieved by treating the flowers with de-mineralized water supplemented with germicides and acidified with citric acid to pH 4.5 to 5.0 but with or without sugar. Hydration is improved when water is de-aerated or acidified or when a wetting agent like Tween 20 at the rate of 0.01 to 0.1 per cent is added. Flower stems are placed in warm water or in a preservative solution in plastic jars at a depth of 2-4 cm and held at room temperature or in cold storage for several hours.

5.6 Impregnation

Sometimes, the cut ends of the flower stems were impregnated for a short time with chemicals. This treatment protect the blockage of the water vessel in the stem by microbial growth and stem decay. Impregnation of cut bases of flowers with high concentration (1000 ppm-1500 ppm) of silver nitrate, nickel chloride or cobalt chloride for 10-15 minutes improves the longevity of several flowers such as aster, gerbera, gladiolus, carnation, chrysanthemum, phalaenopsis, cymbidium and snapdragon.

5.7 Pulsing

The absorption of chemical solutions containing sugars and germicides through the lower cut bases of flower stems is known as pulsing. Pulsing could be used by growers, wholesalers or retail florists in order to enhance the cut flowers subsequent vase life in water. Pulsing is employed with higher concentrations of sugar, mainly sucrose, the percentage of which varies with species and cultivars. Other chemicals used in the pulsing treatments were STS, $AgNO_3$, HQ, MH, AOA, $CaCl_2$, $CoCl_2$, nickel sulphate, aluminium sulphate and benzyladenine. Pulsing is found to be of great value in prolonging life, promoting opening and improving the colour and petal size of petals through osmo-regulation.In *Cymbidium* hybrid 'Red Princess' pulsing with 5 per cent sucrose increased vase life (56 days) followed by sucrose @ 8 per cent (54.78 days). In *Phalaenopsis*, pulsing with 0.5mM STS for 24 hours blocked the deleterious effect of ethylene. In *Dendrobium* cv. 'Sonia' pulsing with 4 per cent sucrose + 400 ppm HQ recorded the highest vase life of 21.33 days. Inflorescences pulsed with 6 per cent sucrose + 400 ppm HQ recorded the highest sugar content in the flowers (27.64 per cent).

5.8 Bud Opening

It is a procedure of harvesting flowers at a stage earlier than normally considered as the cutting stage and then opening the buds off the plant. Such types of postharvest handling could be applied by growers or wholesalers. Bud opening of flowers increases longevity of cut flowers by reducing the sensitivity of flowers to extreme temperatures, low humidity and ethylene, saving space during shipment and extending the useful storage life. The sugar concentration used is lower than the concentration of pulsing and the optimum temperature is kept lower.

In *Dendrobium* hybrids, HQS or $AgNO_3$ (50ppm) was effective for opening of tight bud cut flowers. In *Cymbidium* hyb. 'PCMV', sugar(4 per cent) + salicylic acid (200 ppm) showed maximum per cent of flower opening (75 per cent) and vase life (45 days) followed by sugar (4 per cent) + $Al_2(SO_4)_3$(100 ppm) (57 per cent and 44 days) and sugar (4 per cent) + 8-HQS (200 ppm) (53.8 per cent and 44 days) over control (bud drop and senescence on 27th days), respectively.

5.9 Preservatives

Preservatives are used in the holding solutions in the form of tablets containing a mixture of chemicals such as sugars, germicides, salts, growth regulators etc. Besides, the chemicals are employed during conditioning, pulsing and for making bud opening solutions to improve flower shape, size and opening and colour of the flowers.Sugar, biocide, anti-ethylene compounds and hydrated compounds are used for conditioning. The sugar and biocide solutions are effective for opening of bud cut flowers. The vase solution should contain sugars, acidifying agent and a biocide. Citric acids are mainly used for acidifying agent and hydroxy quinoline as biocide. Metallic salts like silver nitrate, cobalt chloride, aluminium sulphate, zinc sulphate, calcium nitrate and nickel chloride have been found for prolonging postharvest life of various cut flowers. Among several growth regulators used to increase vase life of cut flowers, BA, IAA, NAA, 2,4,5-T, GA3, B9, CCC are common. New chemicals that had been found promising as floral preservatives are ethylene inhibitors like amino-oxyacetic acid, 1-amino cyclopropane, aminotriazole, aminoethoxy vinyl glycine, alpha aminoisobutyric acid, diazocyclopentadiene and phenidone. Different chemicals used in holding solution for improving vase life of orchids are listed in Table 5.

Table 5: Holding Solutions for different Types of Orchids

Name of Orchid	*Holding Solution*
Arachnis, Aranda, Aranthera, Cattleya	STS (1mM) + 1 per cent sucrose
Cymbidium	8-HQC 200 ppm + sucrose 2 per cent
	1-MCP (500ppb)
	1 per cent Sucrose + STS (1 mM)
Dendrobium	8-HQC (200 ppm) + sucrose (2 per cent),
	0.5 mM AOA + 4 per cent sucrose,
	$AgNO_3$ (30 ppm) + 4 per cent sucrose
	400ppm HQ + 30ppm $AgNO_3$ + 2 per cent sucrose
	200 ppm 8-HQS + 50ppm $AgNO_3$ + 8 per cent sucrose
Oncidium	8-HQC (100-200 ppm) + 4 per cent sucrose
	kinetin (50ppm) + 4 per cent sucrose
Paphiopedilum	8-HQC (200 ppm) +2 per cent sucrose
Vanda	$AgNO_3$ (30 ppm) + 1.5 per cent sucrose

Source: De *et al.*, 2013.

5.10 Grading and Packing

5.10.1 Grading

Grading is done based on the basis of parameters like appearance, stage of maturity, blemishes or injuries due to diseases, infestations caused by insect pests, colour and size of the bud, straightness, strength as well as length of stem.Flowers are generally grouped into bunches of 5, 10, 12 or 20 stems and loosely tied with rubber bands. Before placing them in the package, individual flower bunches are wrapped with suitable packing materials like cellophane paper, kraft paper, newspaper, tissue paper or corrugated cardboard sheet. For local markets, the bunches are held in buckets containing water or preservative solution. It is advisable that for long distance transport and storage, flower bunches are held in dry cardboard boxes. The minimum length of boxes should be about twice the width and it's width about twice the height. Use of telescope-style boxes made of CFB is ideal. Grading of cut flowers of orchids is done based upon flower count and spike length of different genera (Table 6).

Table 6: Grading of Cymbidium and Dendrobium Cut Flowers

Cymbidium	*Grade*	*Flower Count*	*Spike Length*	*Other Consideration*
Cymbidium Standard	AAA	>12	1.25m	☆ Strong straight stems, uniform length, no marks on flowers.
	AA	8	90 cm	☆ Bent crooked spikes but with perfect flowers.
Cymbidium Miniature	XL	>15	65+	
	L	12-14	55-64	
	M	8-11	40-54	
	S	< 5	30-39	
		Flowering		
Dendrobium	Small (S)	4-5	30cm	
	Medium (M)	6-8	40cm	
	Large (L)	8-10	45cm	
	Extra Large	>10	50cm	

5.10.2 Packaging in orchids

Cut flowers are inserted in tube containing water or water with preservatives or simply wrapped in wet cotton swab while is covered with a piece of plastic and tied with rubber band to keep it in place. Flower spikes are bunched into bunch of 5 or 10 or so. Bunches or individual spikes are placed inside the box in alternate fashion. Ethylene scrubbers with $KMnO_4$ or Purafil are also be kept in the box. For export purpose, packing of flowers in two piece box is a better option. In *Cymbidium*, single flowers backed by a fern leaf are inserted in small flasks containing preservative solution, the flasks are then packed in 3 sided box with a display window. In *Dendrobium* hybrid 'Sonia-17' a low gauge polyfilm of 100 gauge thickness the cotton dipped in 8-HQS (25ppm) covering the base of the spike rended maximum vase life and flower quality. Number of cut spikes packed per box for standard Cymbidium orchids are given in Table 7.

Table 7: Packing Requirement for Cymbidium Orchids

Stems/Box	Standard Box	Super Box
6	15+	—
8	11-14	14+
10	9-10	11-13
12	7-8	9-10
15	5-6	—

5.11 Storage

Low temperature treatment during storage or shipment period reduces the entire metabolism in the tissues, slows down the respiration, transpiration and ethylene action and retards the multiplication of bacteria and fungi. Storage of cut flowers are presented in Figure 2.

In general, temperate orchids are stored at lower temperature even at 5 °C in cold chambers whereas tropical orchids are stored at 7-10 °C. A 90-95 per cent relative humidity is necessary during storage to minimize moisture loss and to prevent wilting.

There are two types of cold storage methods, namely 'Wet storage' and 'Dry storage'. In wet storage, flowers are stored with their bases dipped in water or preservative solution for a short time. Dry storage method is are used for long term storage. In this method, fresh flowers are harvested in the morning, graded and sealed in plastic bags or boxes to prevent the loss of moisture.

In Controlled Atmosphere (CA) storage, cut flowers are kept in gas tight cool chambers equipped with cooling systems at a higher level of CO_2 and lower level of O_2 to reduce the respiration rate and production and action of ethylene. Generally, the concentration of $CO_2$is maintained higher than 4 per cent and not below 0.4 per cent in CA storage (Table 8).

Table 8: Storage of Orchid Cut Flowers

Name of Orchid	Storage Temperature (°C)	Storage Period (days)
Arachnis, Aranda, Aranthera, Ascocenda, Epidendrum	8-13 °C	10-14
Dendrobium	5-7 °C	10-14
Cymbidium	1-4 °C	14
Oncidium, Phalaenopsis, Odontoglossum, Cattleya	7-10	14
Paphiopedilum	–0.5-3. ºC	20

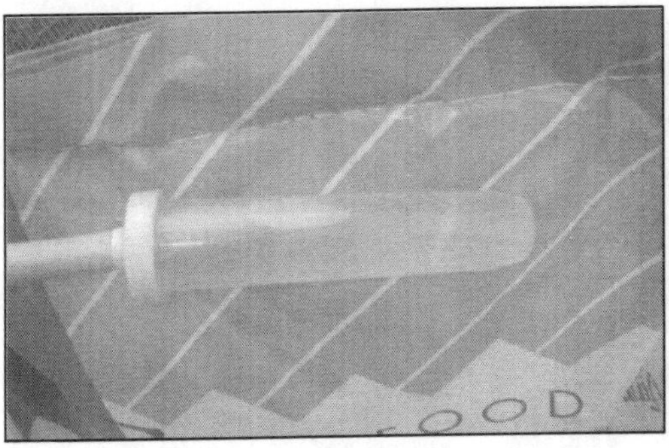

(a) Cut Flowers are Inserted in Tube with Preservatives

(b) Individual Spikes Inside the Box in Alternate Fashion for Domestic Transport

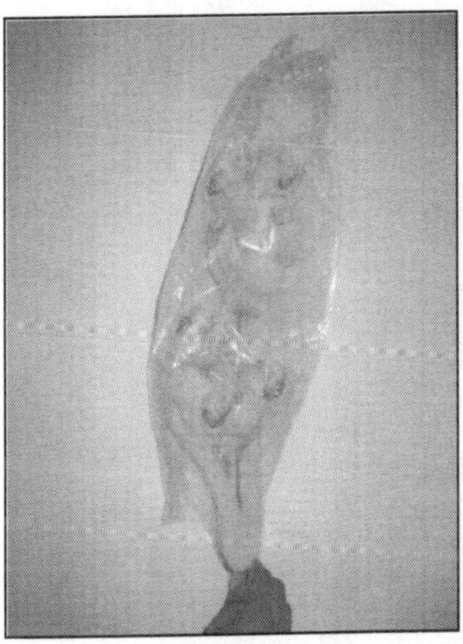

(c) Polyethylene Wrapping of Cut Cymbidium

(d) Packing of Cut Cymbidiums with Local Packaging Materials

(e) Packing of Cut Cymbidiums for Export Markets

Figure 2: Packaging of Cut Flowers

5.12 Transport

Being short lived and perishable in nature, flowers are request to be delivered to destination as early as possible immediately after harvest. For long distance markets, cut flowers are transported by cargo planes, merchant ships and trucks. Other modes of transportation are head loads, bicycles, two-three wheelers, cars, vans etc. Hence, for long distance transportation, advanced methods of postharvest handling like cooling, conditioning, impregnation, pulsing, bud opening and packaging were followed.

Short time pulsing of flowers with optimal concentration of sucrose, $AgNO_3$, STS and growth regulators was important for long term truck and sea shipments. The best method of transport of most of the cut flowers including temperate orchids is under refrigeration from the grower to final consumers.

6.0 Domestic Marketing and Export

India is blessed with varied agro-climatic zones congenial for production of sensitive and delicate floriculture products. According to National Horticulture database 2012 the area under floriculture production in India was 253.65 thousands hectares with a production of 1.652 million tonnes loose flowers and 750.66 million tonnes cut flowers.

The recent export scenario of orchids in Indian context is given in Table 9.

Table 9: Export of Orchids from India

Country	*2010-2011*		*2011-12*		*2012-13*	
	Quantity (MT)	*Value (lakh)*	*Quantity (MT)*	*Value (lakh)*	*Quantity (MT)*	*Value (lakh)*
Bahrain	–	–	–	–	0.20	1.43
Kenya	–	–	0.04	0.05	0.06	0.09
Maldives	0.04	0.18	0.13	0. 56	0.27	0.96
Quatar	–	–	0.01	0.02	–	–
Singapore	4.77	3.08	–	–	–	–
Sri Lanka	–	–	–	–	0.02	0.05
UAE	0.04	0.06	0.20	0.35	–	–

Source: DGCIS Annual report.

India's total export of floriculture was Rs. 423.43 crores in 2012-13. The major importing countries were USA, Netherlands, Germany, United Kingdom, Japan, Canada and Japan. There are more than 300 export-oriented units in India.

The value of fresh cut orchids and buds trade during 2007-2012 with the average trade value was US $ 483 million. In 2012, there were more than 40 exporting orchid countries and 60 importing orchid countries around the world, and the total size of the global trade is US $ 504 million (Table 10).

Table 10: Global Trade Value of Fresh Cut Orchids and Buds (Unit: Million US$)

Year	*2007*	*2008*	*2009*	*2010*	*2011*	*2012*
Import	233,734,023	252,647,645	232,568,129	251,445,523	265,702,077	267,196,847
Export	230,470,421	238,702,950	217, 781, 745	227,389,789	244,996,271	237,543,797
Total	464,204,444	491,350,595	450,349,874	478,835,312	510,698,348	504,740,644

Source: Department of Foreign Trade, Thailand (2013).

The Netherlands is the top most orchid exporting country (39. 67 per cent) followed by Thailand (28.41 per cent), Taiwan (10 per cent), Singapore (10 per cent) and New Zealand (6 per cent), respectively. Importing countries are mainly Japan (30 per cent), UK (12 per cent), Italy (10 per cent), France (7 per cent) and the USA (6 per cent), respectively. The total orchid cut flower trade of the world mostly consists of 85 per cent Dendrobium species and 15 per cent Phalaenopsis and Cymbidium species and Asia is the main source of orchid to enter the world. Major markets in Asia are occupied by Japan and Singapore. Total imports of orchids by Japan accounted for US$ 57.4mn in 2008 making it the largest importer of the orchids in the world. The main sources of imports include Thailand, Taiwan, New Zealand and Malayasia together accounting for as much as 96.5 per cent of the total imports of the orchids by Japan in 2008. Imports by Singapore of fresh orchids are mainly for Malayasia, Thailand and Taiwan Indian exports of orchids are insignificant in global trade and this clearly indicates that there is vast scope of increasing India's exports to Singapore particularly considering the proximity of the country and India's East Policy.

Thailand is the largest world exporters of tropical orchids. China is the largest consumer of orchid cut flowers from Thailand followed by Japan, USA, Italy, India, Taiwan, Vietnam and the Netherlands. Other significant orchid genera being exported from Thailand were *Mokara,Aranthes, Aranda, Oncidium, Vanda, Arachnis* ans *Ascocenda* with very small market.

In India, it has take almost a century for Sikkim and the northeastern region to realize the potential of *Cymbidium*, with *Dendrobium, Vanda, Cattleya, Phalaenopsis* and *Oncidium* orchids as revenue earners. While a late start it is turning out to be a promising sector. Sikkim now annually produces Rs 20 million worth of orchids on an average. A cut stalk (spike) costs between Rs.20 and Rs.150 depending on the quality.The state government has introduced a special package in 18 clusters under which 50 flori-farmers have been given 500 orchid plantings (saplings) each with necessary support for cultivation, training and marketing. The state has to overcome several challenges to send the produce out to the national market. The first hurdle is transportation to the nearest market place. An effective cold chain is another area of concern.

Orchids and other horticultural products are marketed by SIMFED - the government's national sales network. Veteran flower trader Andy Warren, managing director of the New Zealand-based company Bloomz, proposed that Sikkim and the northeastern states - the country's traditional orchid hothouses - should concentrate on the domestic market and explore its "optimal capacity" for high returns instead of eyeing bigger export share globally.This region lacks postharvest facilities like good packaging houses and cold storages. During hot summer logistics to move the harvested products are inadequate. Flowers like orchids wilt. The region needs to look at expanding markets in New Delhi, Mumbai, Bangalore and Chennai to fetch good prices. The critical thing is to ensure top quality in the market every day, Warren, who has been trading flowers with India for the last 20 years, told IANS at the Sikkim International Flower Show 2013. According to his concern, going to the international market is a "whole new ball game" with its stringent set of standards.

7.0 Conclusion

The orchids have taken a significant position in cut flower industry due to this attractiveness, long shelf life, high productivity, right season of bloom, easy packing and transportation. These account for a large share of global floriculture trade both as cut flowers and as potted plants and are estimated around 10 per cent of international fresh cut flower trade. Keeping in view of the export potential of Orchids, and based on the recommendations of the planning commission the Indian Council of Agricultural Research established a National Research Centre for Orchids in 1996.Orchids are very high value floriculture produce. Integration of crop management with postharvest management is essential for obtaining success for this industry. Selection of appropriate variety/species suitable for the location along with establishment of appropriate postharvest management facilities for handling, packaging, storage and transport are the key requirement for any successful venture on orchid industry. For orchard trade to be successful attention need to be given availability of adequate planting materials and quality flower spikes to compete in international markets. There is also immediate need to upscale production strategies. The qualitative aspects need to be given due emphasis for value realization in the market. For this, maximum growing standards need to meet requirements, market standards and consumer preferences. However, erect stems, freedom from pests and mechanical injury, flower freshness are some of the indispensable criteria.

8.0 Important References

Bhattacharjee, S.K. 1997. Packaging fresh cut flowers. *Indian Horticulture*, **41** (4): 23-27.

Bhattacharjee, S.K. and De, L. C. 2005. *Postharvest Technology of Flowers and Ornamental Plants.* Pointer Publishers, Jaipur, Rajasthan (India), Pp. 440.

Chowdhery, H.J. 2001. Orchid diversity in north-east India. *J.Orchid Soc. India*, **15**: 1-17.

Chowdhery, H.J. 2009.Orchid diversity in northeastern states of India. *J. Orchid Soc. India*, **23**(1-2): 19-42.

De, L.C., Deb, P. Chhetri, Geetamani, Medhi, R.P. and Pokhrel, H. 2013. *Postharvest management in orchids.*Technical Bulletin No.16, NRC for Orchids, Pakyong, Sikkim, pp. 37.

De, L.C. and Debnath, N.G., 2011. NRC for Orchids, *Vision 2030*. Pakyong, Sikkim, India, Pp.28.

De, L.C. and Bhattacharjee, S.K. 2000. Methods for prolonging vase life of cut flowers- a review. *Orissa Journal of Horticulture*, **28**(1): 73-78.

De, L.C. and Bhattacharjee, S.K. 2002. Vase life of cut rose 'Queen Elizabeth' as affected by aquatic fungi and bacteria. *Indian Rose Annual*, **18**: 86-90.

Hajra, P.K. and De, Aparajita. 2010. Phyto-geographic analysis of orchid flora in India, *J. Orchid Soc. India*, **24** (1/2): 43-46.

Jawaharlal, M., Dinesh Babu, M.K. and Indhumathi, 2006. Postharvest packaging techniques for Dendrobium Hybrid Sonia -17. *J. Orn. Hort.* **9** (1): 16-19.

Jomy, T.G. and Sabina,G.T. 2002. Effect of conditioning and pulsing on vase life of Dendrobium Sonia inflorescences. *J. Orn. Hort.*, **5**(1): 80-81.

Sarkar, Indrajit, Mandal,T. Naveen Kumar, P., Kumar, Rajiv, Misra, Sanyat, Misra, R.L. and Singh, Kishan Pal,2009. *Temperate orchids*. AICRP on Floriculture, Technical Bulletin No.**28**, IARI, Pusa, New Delhi-12, Pp.77.

Singh, Foja 1990. Indian orchids. *Indian Horticulture*, **35**(1): 14-15.

Chapter 35
Potato

R. Ezekiel

Former Head, Division of Crop Physiology, Biochemistry and Postharvest Technology, ICAR-Central Potato Research Institute, Shimla – 171 001, H.P., India

E-mail: rajarathnamezekiel@yahoo.co.in

1.0 Introduction

The importance of potato (*Solanum tuberosum* L.) as food is well recognized throughout the world. It originated in South America, where it was cultivated by Incas. It is believed that Potato was introduced into India by the Portuguese in the 17th Century. However, it was the English who were responsible for its wide spread cultivation in the hills of the country. The potato is a staple food in Europe and North America. But, in India, potato is still considered a vegetable and is given the pride of place amongst vegetable crops. Earlier potato was considered as a crop of developed nations in the world. This is no longer the case and in fact developing countries now produce about 35 per cent of the world's potatoes and production in developing countries is growing rapidly. India has emerged as the second largest producer of potatoes in the world after China. Today potato is being grown almost throughout the country and under diverse agro-climatic conditions. The potato growing areas in the country can be broadly classified into three zones *i.e.* hills, plains and plateau. About 90 per cent of the potato crop is grown in the plains under sub-tropical conditions. Potato tuber is a wholesome food with plenty of carbohydrates and good quality protein of high biological value. Besides, potatoes contain vitamins, minerals and anti-oxidants. Potatoes are rich in vitamin C and potassium and low in sodium.

There are several success stories in Indian Agriculture, which have contributed to the present position of plenty and the success story of potato in India should easily rank among the best. During the last sixty years, there has been a revolution in potato production in India. The crop once considered best suited to temperate conditions has emerged as one of the major food crops in India and now is well suited to the subtropical Indian conditions.

2.0 Status of Cultivation

In 1949-50, the potato production in the country was 1.54 million t, the area 0.234 million ha and the average yield 6.58 t/ha. After sixty years *i.e.* in 2009-10, the production has increased to 37.3 million t, the area to 1.83 million ha and the average yield to 20.38 t/ha. Now, India is the second largest potato producer in the world. About 90 per cent of the potato crop in India is cultivated in the plains during the winter season. The Indo-Gangetic region comprising Uttar Pradesh, Bihar and West Bengal accounts for 80 per cent of the potatoes grown in the plains. Uttar Pradesh is the largest potato producing state in India followed by West Bengal and Bihar. The potatoes are harvested in this region during January-march. State-wise area, production and productivity of potato in India during 2012-13 are presented in Table 1.

Table 1: Area, Production and Productivity of Potato in India 2012-13

Sl.No.	*State*	*Area (thousand ha)*	*Production (thousand t)*	*Productivity t/ha*
1.	Assam	99.77	975.27	9.8
2.	Bihar	322.46	6640.55	20.6
3.	Gujarat	81.27	2499.73	30.8
4.	Haryana	29.47	676.02	22.9
5.	Jharkhand	47.21	659.61	14.0
6.	Karnataka	44.40	698.30	15.7
7.	Madhya Pradesh	108.87	2299.00	21.1
8.	Punjab	85.25	2132.31	25.0
9.	Uttar Pradesh	803.76	14430.28	23.9
10.	West Bengal	386.31	11591.30	30.0
11.	Others	183.10	2741.20	15.0
	Total	1992.20	45343.60	22.8

Source: Indian Horticulture Database, NHB 2013.

3.0 Commercial Cultivars

In India several commercial varieties of potato are grown in different states. Some of the most common varieties with their special characters are mentioned below (Figure 1).

a) Kufri Badshah

This is a hybrid between Kufri Jyoti and Kufri Alankar. The tubers are white, large, oval with smooth skin, fleet eyes and dull white flesh. Tubers turn pale purple on exposure to light. The tubers mature between 100-110 days with very high yield potentiality of 50 t/ha. The tubers have medium dormancy. This is commercially cultivated in parts of Gujarat, Haryana, Jammu and Kashmir, Madhya Pradesh, Punjab and Uttar Pradesh.

b) Kufri Chandramukhi

The tubers are slightly elongated shape having medium dormancy of 6-8 weeks. It matures in about 90-100 days after planting having yield potentiality of 30 t/ha in plains. This is commercially cultivated in North and South Indian hills, parts of Bihar, Gujarat, Karnataka, Madhya Pradesh, Maharashtra, Punjab, Uttar Pradesh and West Bengal (Figure 1a).

c) Kufri Chipsona-I

Tubers are white, medium to large with smooth skin, fleet eyes, dull white flesh. The variety is medium in maturity (90-110 days), with very high yield potential of 40 t/ha and is resistant to late blight disease. This is commercially grown in Bihar and U.P. Due to high dry matter and low reducing sugars, the variety is suitable for making chips (Figure 1b).

d) Kufri Jyoti

Tubers are white, large, oval with smooth skin, fleet eyes, white flesh. There is tendency to crack. The variery has high yield potential upto 30 t/ha in the plains with maturity at 90-100 days. The variety is having medium dormancy of 6-8 weeks. This is commercially cultivated in North and South Indian hills, parts of Bihar, Gujarat, Karnataka, Madhya Pradesh, Maharashtra, Punjab, Uttar Pradesh and West Bengal (Figure 1c).

e) Kufri Lauvkar

This is a hybrid of cross between Adina and Sarkov. It can be grown both in Kharif and Rabi seasons in the plateau regions of Karnataka, Madhya Pradesh and Maharashtra. This is an early variety and matures in 75-80 days with yield potentiality of 30 t/ha. The tubers have medium dormancy of 6-8 weeks.

f) Kufri Pukhraj

The tubers are white, large, oval, slightly with smooth skin, fleet eyes, yellow flesh.This is an early maturing (75-90 days) with very high yield potential of 40 t/ha and moderately resistant to early and late blight diseases. Tubers have medium dormancy of 6-8 weeks and are not suitable for processing (Figure 1d).

Record levels of production during the last two decades have resulted in several postharvest problems and the major problem is that of storage. About 80 per cent of the total potato production comes from the Indo-Gangetic plains, which is harvested in February-March. Therefore, the arrivals in the market during this period are huge creating gluts resulting in distress sale by the farmer. Storage helps in regulating the arrival in the market and reduces transport bottlenecks at the peak period of production. Since fresh potatoes are available only for 2-3 months in a year, the

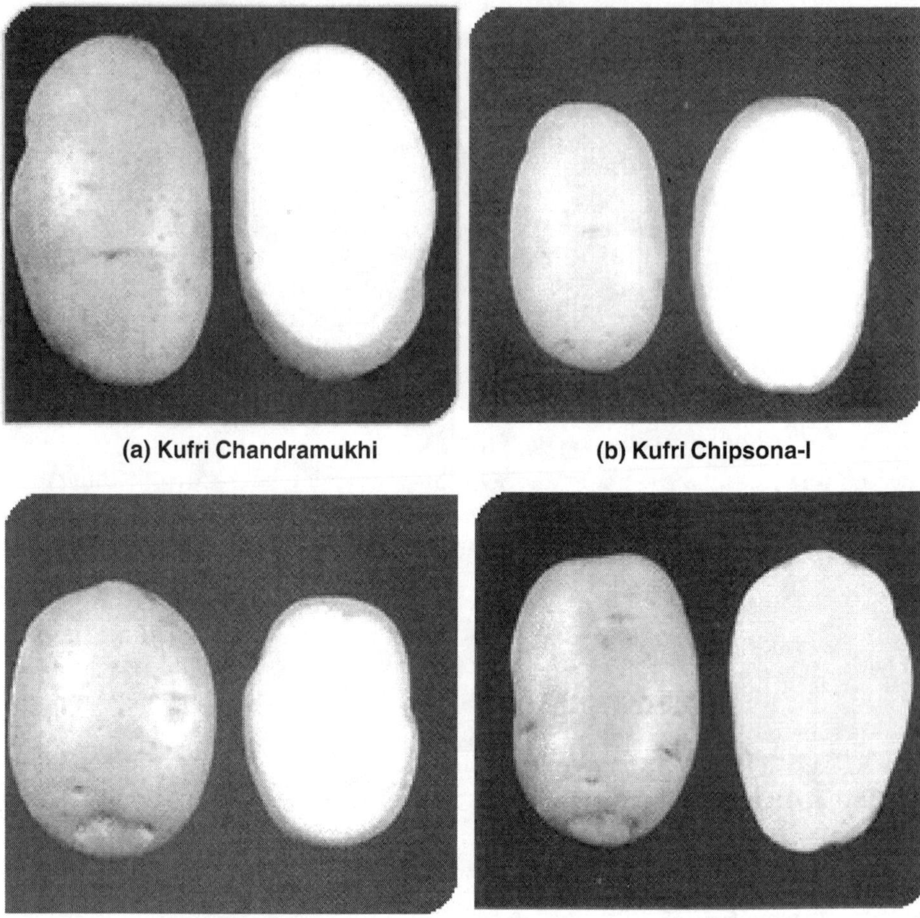

(a) Kufri Chandramukhi (b) Kufri Chipsona-I

(c) Kufri Jyoti (d) Kufri Pukhraj

Figure 1: Commercial Cultivars of Potato

requirement of table potatoes for the remaining period of 9-10 months have to be met from stored potatoes. Potatoes being semi-perishable in nature with 80 per cent water content cannot be stored for 9-10 months without refrigeration. Therefore, cold storage becomes essential.

The high water content in potatoes makes handling and storage difficult. It has been estimated that under tropical and sub-tropical conditions, losses due to poor handling and storage can amount up to 40 per cent. Therefore, it is of utmost importance to minimize storage losses. It makes sense to minimize storage losses to increase the availability of potatoes as it costs less to store than to produce a given quantity.

4.0 Nature and Causes of Postharvest Losses

Good postharvest management begins at the production stage itself. The good quality of the tuber is greatly influenced by the cultural practices. Potatoes grown on improperly prepared fields are damaged while harvesting mechanically and have reduced sugar quality. Conditions for high yield and for good storage characteristics may be contradictory. Further, adequate pest and disease management is essential for producing potatoes with good storage quality. Therefore, proper management of pre-storage factors that affect the keeping quality of potatoes is the first step for good storage management. Whatever may be the storage method, its success depends on the quality of the tubers entering storage. If the tubers entering storage are not sound, the tubers that come out of storage will also be of poor quality.

4.1 Dehaulming

Removal of haulms at full maturity (*i.e.* 90-110 days after planting, depending upon the variety) is essential for proper curing of skin of tubers. Dehaulming is done manually by cutting with a sickle close to the ground or by spraying non-selective herbicide paraquat @ 0.5kg/ha. If haulms are to be removed manually, it is preferred to keep the vines/haulms on the ridges to protect exposed tubers from high temperature and direct sunlight. Re-growth of leaves if any should also be cut after a week of dehaulming.

4.2 Time of Harvesting

Potatoes should be harvested in dry weather. Irrigation should be stopped two weeks before dehaulming. Potatoes harvested under wet soil conditions must be dried before storage because even little moisture on the surface of tubers could lead to infection and rotting during storage. Only mature tubers should be stored. Immature tubers have poor keeping quality due to lower dry matter content and a weak skin (Figures 2a and b). Harvesting is done 10-15 days after haulm cutting to facilitate proper skin curing. In spite of best efforts, cutting and bruising cannot be completely eliminated while harvesting and therefore, curing is essential to heal the wounds. Suberization is the process by which wounds are healed in potatoes and the optimum conditions for suberization are a temperature of around 25 °C and a relative humidity of 95 per cent.It is important to grade the produce into different grades before storage and grading should be done according to the requirement. Green, cut/cracked, damaged and under size tubers (less than 2.5 inches) should be removed and only well graded, defect free tubers of desired size should be stored.

Figure 2: (a) Immature Tubers and (b) Well Cured Mature Tubers

4.3 Dormancy

Dormancy is a state in which tubers will not sprout even when placed under conditions ideal for sprouting (18-20 °C, 90 per cent RH and darkness). All tubers remain dormant for some time and the duration of dormancy varies with the variety. Duration of dormancy is normally counted from the date of harvest. Another method of determining dormancy is from the date of detaching the tubers from the plant *i.e.* dehaulming. This is a better and reliable method as the time between dehaulming and harvesting can vary. Dormancy is expressed as number of days between the date of dehaulming and the date on which 80 per cent tubers are sprouted. A tuber is counted as sprouted if it has at least one sprout longer than 2 mm.

4.3.1 Factors Affecting Dormancy

a) Cultivar

Based on the dormancy period from the date of dehaulming in the plains, Indian potato cultivars can be divided into three categories.

Short dormancy: < 8 weeks, *e.g.* Kufri Bahar.

Medium dormancy: 8-10 weeks, *e.g.* Kufri Jyoti.

Long dormancy: > 10 weeks, *e.g.* Kufri Sindhuri

Even within a cultivar, dormancy can be affected by growth conditions and tuber weight.

b) Growth Conditions

Soil and environmental conditions during crop growth have a strong influence on the duration of dormancy. Cold and wet weather is known to increase the dormancy while dry and warm weather reduces it. Potatoes from crop raised under short-day conditions (plains) have shorter dormancy than those from long-day crop (hills). For example, tuber dormancy in Kufri Jyoti, Kufri Chipsona-1 and Kufri Chipsona-2 was 55,47 and 40 days respectively when grown in the northern plains under short days and 126, 84 and 91 days respectively when grown in the northern hills under long days. Season to season variation in the duration of dormancy can also be considerable due to variation in the environmental conditions during the crop growth.

c) Storage Environment

Storage temperature has a strong influence on the dormancy. Higher storage temperature hastens dormancy release, while storage at a temperature of 4 °C and below prolongs dormancy by preventing sprout growth. Exposure to light during storage also affects dormancy but its effect is less as compared to temperature effect. RH in the stores has very little effect on tuber dormancy. Concentration of gases in the storage atmosphere affects dormancy of stored potatoes. Lower concentration of oxygen (O_2) and higher concentration of carbon dioxide (CO_2) hastens the release of dormancy and stimulates sprout growth. CO_2 concentration in the storage atmosphere is reported to affect dormancy duration and sprout emergence. It is also known to affect the number and length of sprouts in tubers stored at 4-10 °C. Work done at CPRI has shown that exposure of tubers of Kufri Chandramukhi and Kufri Jyoti to 5, 10, 15 and 20 per cent CO_2 concentrations for 7 days reduced the dormancy duration, and 20 per cent CO_2 concentration showed maximum effect, reducing the dormancy duration by 20 days in tubers stored at 18 °C and 90-95 per cent RH. Sprout growth was also stimulated. Longer sprouts were observed when stored at 15 per cent than at 20 per cent CO_2 concentration.

Postharvest losses lead to reduction in the quantity as well as quality of potatoes. Quantitative losses are apparent and attempts are made to reduce these losses whereas quali-tative losses are not apparent and their importance is underestimated. Since qualitative losses can greatly reduce the value of potatoes, adequate attention should be paid to prevent loss in quality as well. Postharvest losses result from physiological or pathological causes.

4.4 Physiological Losses

Storage losses are due to (i) respiration, (ii) transpiration, (iii) water loss, (iv) sprouting and (v) rotting. Potatoes being living material respire and excess of respiration results not only in the loss of dry matter but also in the generation of metabolic heat. Physiological losses occur through natural respiration, transpiration (evaporative loss of water). Major weight loss during storage is due to evaporative loss of water. Water loss occurs largely through periderm and to a smaller extent through lenticels. When the loss of water from tubers exceeds 10 percent, the market value of potatoes is greatly reduced because of shriveling. Freshly harvested immature potatoes loose more water than mature tubers because immature skin is more permeable to water. Sprouting results in increased water loss, as the surface of sprouts is 100 times more permeable to water vapour than the periderm of tuber per unit area and time. Sprouting also results in increased respiration leading to higher dry matter loss. The magnitude of these losses depends largely on the environmental conditions. Two important storage environmental factors that affect the storage behaviour of potatoes are temperature and relative humidity.

4.4.1 Effect of Temperature

Physiological damage can occur from exposure to high or low temperatures both before and during storage. Overheating of tubers either due to direct exposure to sun light or during high temperature and non-refrigerated storage could lead to black heart symptoms. At high temperature storage, the high rate of respiration leads to larger oxygen requirement, which results in asphyxiation. The discolouration and breakdown of inner tissues of tubers is a result of asphyxiation. Tubers exposed to freezing temperatures also suffer internal damage. Tuber respiration during storage results in loss of dry matter. In addition, the heat generated during respiration called respiratory heat also influences the storage temperature and in turn affects the storage life of potatoes.

4.4.2 Effect of Relative Humidity

The relative humidity in the storage atmosphere affects the weight loss of tubers since water loss is directly proportional to the difference in water vapour pressure between the potatoes and the surrounding air. Generally a higher relative humidity of 90-95 per cent is maintained in cold stores to minimize weight loss. The relative humidity in the storage atmosphere also affects the rate of sprout growth and the form of sprouts. The degree of branching in sprouts is greater under lower humidity levels. Relative humidity also affects wound healing. Wound periderm formation is rapid at higher humidity. For example, at a storage temperature of 10 °C, wound healing is faster when stored above 80 per cent RH.

4.5 Pathological Losses

Postharvest losses in potatoes caused by pathogens are greater than the losses due to physiological causes. Physical damage to tubers during harvesting and handling predispose tubers to attack by bacteria and fungi leading to quantitative loss. The more common storage diseases caused by fungi are late blight, dry rot and pink rot. The most severe bacterial disease that causes rotting is soft rot. When infection occurs in the field, rotting begins in the field and continues during storage, for example, late blight, brown rot and pink rot. When infection occurs after harvesting, it is generally through mechanical injury as in the case of dry rot (Figure 6). High humidity and condensation of water on tuber surface can lead to infection by soft rot causing bacteria *Erwinia* spp. Qualitative losses are caused by diseases such as common scab, powdery scab, black scurf and wart, which affect the appearance of tuber and thus reduce the market value of potatoes. Among the insect pests, tuber moth causes maximum damage during storage and is common in potatoes stored under higher temperatures, as is the case with non-refrigerated storage. The larval damage results in weight loss and tuber moth infection greatly reduces market value of tubers.

5.0 Postharvest Management

5.1 Storage Methods

Unlike in the temperate regions of the world where potatoes are harvested and stored during the cold, in India potatoes are harvested in February-March in the Indo-Gangetic plains when the temperatures begin to rise and have to be stored during the hot summer months. Potato storage methods being practised in India can be broadly divided into two categories, *viz.* Short-term storage (non-refrigerated) and Long-term storage (refrigerated).

5.1.1 Short-term Storage

When potato tubers are dormant it is relatively easy to store them under non-refrigerated conditions. But once dormancy is released and sprout growth begins, storage under non-refrigerated conditions becomes difficult as storage losses increase. Though refrigerated storage is the best option for long-term storage needs, considering the rising cost of construction, equipment and energy required for additional refrigerated storage capacity, alternate storage methods should be considered for meeting short-term storage requirements. Further seasonal production pattern, limited alternative market outlets and high cost of cold storage often result in gluts and the consequent reduction in price during the harvesting period. Potato prices rise rapidly in April-May and are almost double in July-August. Therefore, holding the produce for 3-4 months and then selling it in the market will be advantageous. The following non-refrigerated storage systems are effective for storing potatoes for 3-4 months under hot summer conditions in India.

For short-term storage, on-farm storage is economical because it requires only a small investment and no energy. Potatoes need to be stored on-farm for various reasons such as a farmer might decide to hold his produce for a few months, when the prices are low and are likely to increase in a few months time. Storage is economical only if there is an increase in price during the storage period and the increase in price is sufficient at least to cover storage costs. Crops have been stored on-farm for centuries and the storage methods have been varied in different societies. In the first half of this century, potatoes were stored in West Bengal, Bihar and Uttar Pradesh in cool dry rooms or in pits. Potatoes were also stored in heaps and pits in Mumbai and Pune districts of Maharashtra. Potatoes were stored in pits in the erstwhile Bombay state from February to June. Tubers were examined periodically and rotted ones were removed. Table potatoes were disposed off before they started to sprout but seed potatoes were desprouted repeatedly and stored until the planting time. Some of these methods have survived and are practised even today in villages. The traditional methods of storing potatoes are very useful for storing small quantity of potatoes (15-20 tonnes) for 3-4 months due to the following advantages: (i) These methods are very cheap and no investment is needed on storage structures. (ii) The material required are locally available and farmers can do it themselves thus saving on the cost of labour. (iii) They are readily acceptable to farmers. The traditional methods may be less efficient but if their efficiency is improved by minor modifications, the storage losses can be brought down to acceptable limits and they can also prove to be effective on-farm storage methods for meeting short-term storage requirements. Some of the common non-refrigerated methods used for on-farm storage of potatoes are discussed below.

a) *In-situ* Storage

In-situ storage means delaying the harvest of the crop until it is required. The disadvantage is that the crop may be exposed to pest and disease attack and the land remains occupied, and as a result a new crop cannot be planted there. *In-situ* storage of potatoes for up to a period of three months is practised in Meghalaya and Kumaon hills.

b) Sand Storage

Some of the earliest research attempts at proper storage of potatoes were made in the beginning of 1900's and they were aimed at controlling tuber moth which threatened the potato industry in the country. It was discovered by Lafroy and Evans in 1910 that potatoes could be stored in sand and this became a general practice in Bihar. It was also realized that rotting due to bacterial infection should be reduced and for this storage of potatoes free of injury and treatment with 2 per cent copper sulphate solution for 30 minutes before storage was recommended. Potatoes were stored in sand underground and above ground in Punjab during 1919-20. This kind of storage is practised by farmers in Bihar for the storage of small quantity (up to 200 kg) of potatoes meant for home consumption and the storage period can be extended up to few months. A layer of sand is spread on the katcha floor in a corner of the house or below the cot and potatoes are spread on them. They are turned over once in 15 days and the rotted tubers are removed. A total loss of 20 per cent has been observed.

c) Heap Storage

This is a common and popular traditional method of on-farm storage. Heaps have been used successfully for storing potatoes on-farm for 3-4 months. This method of storage is commonly practised in Madhya Pradesh, Uttar Pradesh, Gujarat, Maharashtra and Karnataka. In Ujjain district of MP, heaps are made under the shade of trees (Figure 3) and the quantity stored can vary from 5 to 33 t and the storage duration varies from 2 to 4 months. Smaller heaps are circular in shape with a diameter of 5 to 7 m. The height varies from 1 to 1.5 m. Bigger heaps are 3 to 6 m wide with length varying depending upon the quantity of potatoes stored. The heaps are covered with 30 to 50 cm thick layer of straw or dried sugarcane leaves. The temperature inside the heap is around 20 °C at the time of making the heap and it increases to 30 °C by the end of the storage period. The temperature inside the heap is 11 to 14 °C lower than the ambient and the notable thing about heaps is that the decrease in temperature is achieved without any cooling mechanism. The RH inside the heap ranges from 46 to 63 per cent and this is 19 to 20 per cent higher than the ambient conditions. In Deesa district of Gujarat, 5 to 40 t of potatoes are stored for 2 to 3 months. While potatoes are stored in bigger heaps for 3 to 4 months in MP and Gujarat, in Karnataka smaller heaps of 1 to 20 t are made and potatoes are stored for 1 to 2 months in districts of Bangalore, Hassan and Dharwad. Heap is locally called *thamboo*. Storage losses vary from place to place (Table 2) and are dependent on temperature at the time of storage, duration of storage, variety, thickness of the straw covering layer and protection from rain and direct sunlight.

Weight loss forms a major part of total losses in heaps. The storage losses of about 10 per cent are considered acceptable and more than 10 per cent is considered unacceptably high. Furthermore, when weight loss exceeds 10 per cent, it leads to shriveling of potatoes resulting in drastic deterioration in quality. Therefore, it is necessary to improve the efficiency of heap method of storage and bring down the storage losses to less than 10 per cent. Work done at Central Potato Research Institute has shown that it is possible to achieve this by erecting a thatched roof over the heap and by advancing the date of storage from mid-March to end of February (Table 3). A thatched roof to cover heap helped in reducing the temperature by 2 °C besides giving protection from the rain. Advancing the date of loading of potatoes into heaps made the maximum difference. At Jalandhar and Modipuram, when potatoes were stored before 1st of March, storage losses were minimum. An additional advantage is that potatoes stored in heaps have lower reducing sugar content, which range from 75-131 mg/100g tuber (a reducing sugar level of up to 150 mg/100g is considered acceptable for chipping) and therefore, produce highly acceptable chips with colour scores of 1-3. (Chip colour was scored on a scale of 1-10 of increasing darkness and a chip colour score up to 4 was considered acceptable, and more than 4 was unacceptable).

Table 2: Losses of Potatoes during on-farm Storage in Heaps at different Locations in the Country

Location	*Variety*	*Storage Period*	
		(Days)	*Total Loss (per cent)*
Deesa (Gujarat)	Kufri Jyoti	90	32.8
	Kufri Badshah	90	26.0 to 37.4
Hassan (Karnataka)	Kufri Jyoti	90	19.5 to 32.0
Jalandhar (Punjab)	Kufri Jyoti	105	9.9 to 12.1
	Kufri Badshah	105	12.7 to 19.5
Modipuram (UP)	Kufri Jyoti	90	7.4 to 14.7
	Kufri Bahar	90	9.5 to 11.2
Patna (Bihar)	Kufri Jyoti	75	8.4 to 31.3
	Kufri Sindhuri	75	34.3 to 36.7
Ujjain (MP)	Kufri Lauvkar	120	12.9

Table 3: Reduction in Storage Losses of Potatoes (Kufri Bahar) Stored in a Heap under different Conditions for 90 Days

Storage Loss (per cent)	*Heap with Vent in the Sun*	*Heap in the Sun*	*Heap under the Shade of a Tree*	*Heap Under the Shade of a Tree and Protected with a Thatched Roof*
Sprout weight*	2.6	2.9	2.6	1.2
Rottage	6.0	5.4	4.6	0.5
Weight loss	15.6	9.4	9.2	8.1
Total losses	24.2	17.7	16.4	9.8

* Sprout weight in g/100g tuber FW.

d) Pit Storage

Pit is another traditional method of on-farm storage. Potatoes were stored in pits in the erstwhile Bombay state from February to June. In Meghalaya and Uttaranchal also, seed potatoes are stored in pits. Pits have been used for storage of potatoes for 3 to 4 months in Ujjain district of Madhya Pradesh. Two types of pits are used for the storage of potatoes and they are *katcha* pit and *pucca* pit. *Katcha* pit is a simple pit dug under the shade of trees. It is generally rectangular in shape measuring 6 m in length, 4.5 m in width and 3 to 4 m in height. About 10-30 t of potatoes are stored in each *katcha* pit.Now-a-days use of *katcha* pit for storing potatoes has decreased mainly due to seepage of rainwater into pits leading to higher storage losses. *Pucca* pit is different from *katcha* pit in that it is lined with a layer of bricks made in the form of a circular well and is generally bigger in size (Figure 4).

Furthermore, it has a longer life, without recurring expenditure on digging and maintenance. A typical *pucca* pit measures 4.5 m in diameter and 4 m in depth and the quantity of potatoes, which can be stored varies from 10 to 40t. The temperature inside the pit at the time of storage is 19 °C and may increase to 29 °C by the end of the storage period. The temperature inside the pit is 11 to 15 °C lower than the ambient and the RH inside the pit ranges from 52 to 73 per cent and this is 25 to 33 per cent higher than the ambient. The storage losses of 10 to 40 per cent in pits are unacceptably high. It is necessary to improve the efficiency of pit method of storage to bring down the storage losses. A major part of total losses is due to rotting and the high incidence of rotting is caused by the pressure exerted by the top layers on the bottom layers of potatoes. By placing a platform made of bamboos placed 2 inches apart and 2 feet above the base of the pit and by reducing the quantity of potatoes stored, it is possible to bring down the storage losses to less than 10 per cent after 3 to 4 months of storage. Like in heap, potatoes stored in pit have lower reducing sugar content, which ranges from 136-148 mg/100g tuber FW and, therefore, produce highly acceptable light colour chips with colour scores of 1-3. The storage cost comes around Rs.18/q.

e) Room Storage

Traditional methods of potato storage were in vogue in the beginning of 1900's and storage in a cool dry room was one of the most common traditional methods of potato storage. Considerable quantities of seed potatoes were stored in cool dry rooms in the plains of Bihar, Uttar Pradesh and West Bengal during February-October. It is common to store potatoes in a dark, cool room in the basement or in a corner of the house for 2 to 3 months. In Himachal Pradesh, Meghalaya and Uttarakhand, seed potatoes are either heaped or stored in gunny bags or bamboo baskets, in such rooms. In the plains, potatoes are stored in rooms in U.P. and M.P. Though the storage losses are as high as 27 to 46 per cent after 3 months due to excessive sprout growth, farmers in northern plains continue to store potatoes this way because of the higher price fetched in June-July. In the 1970's storage of potatoes in ordinary *katcha* farm store and double walled *katcha* farm store were tried at CPRS, Patna.

f) Ventilated Wooden/Bamboo Structures

Storage in diffused light helps in retarding sprout growth of potatoes. Accordingly seed potatoes were stored in diffused light in Assam during 1931-32. Small wooden/bamboo structures are built on a slightly raised site and potatoes are stored in these structures. Wooden planks are used in some places like Meghalaya while in Assam, bamboo is used. These structures are made above the ground level supported by wooden pillars and are well aerated. In Assam, bamboo structures are made in such a way that diffused light enters the store so that the tubers turn green and have small, stout sprouts (Figure 5). These bamboo structures have been found useful for on-farm storage of seed potatoes. The quantity of seed potatoes stored in each structure varies from 0.2 to 1 t. After 7-8 months of storage, 2 to 10 per cent losses occur due to rotting and the weight loss varies from 20 to 35 per cent.Work done at CPRS, Kufri has shown that seed potatoes can be stored successfully in a diffused light store in the hills of Himachal Pradesh.

g) Non-refrigerated Air-Cooled Store

Potatoes can be stored in non-refrigerated air-cooled stores in the hills. The cooler temperatures at night are made use to cool the store. The ventilators of the insulated store are kept open during the night to draw in the fresh cool air and are closed during the day to prevent the entry of warm air. This has been tried at the Central Potato Research Station, Muthorai in the Nilgiri hills for the storage of potatoes.

h) Insulated Ventilated Cool Store

An insulated ventilated cool store was constructed at Daurala farm of Central Potato Research Institute Campus (CPRIC), Modipuram by the Central Potato Research Institute (CPRI) in collaboration with the Central Warehousing Corporation of India Ltd. in 1977. The store is located in a grove of trees. It is a masonry structure measuring 9.15m x 4.57m with 13 air-inlets and 22 ventilators, each 0.91 m x 0.30m on the walls. The air-inlets and ventilators each 0.91m x 0.30m and 2.13 m above floor level are provided with wooden shutters to exclude light from the store. A door 1.83m x 0.22m (with an L-shaped break-wall having an opening towards the east) is provided to serve as an ante-chamber. The ceiling walls, door, ventilators and air-inlets have been insulated with fibreglass. A coating of bitumen has been used on the wall to act as a vapor barrier. The roof consists of two parts, a shorter and steeper slope and a longer but less steep northern slope. The former is 1 m higher than the latter at the ridge. A north-facing row of sky-light ventilators is also provided. The parapet wall on the western side is raised to act as a screen. Store walls are white washed and the roof painted with aluminium paint to reflect sun-light. To control ventilation, two exhaust fans are provided, one each on the eastern and the western walls. Thus the store incorporates all the principles of durability, insulation and ventilation as well as other ancillary devices to reduce the heat load. In 1978, the daily minimum temperature in the cool store ranged between 8 °C in the beginning of March to 22 °C in the beginning of June, and the maximum temperature between 14 °C to 24 °C.During this period, the corresponding range of maximum temperature was 8 °C to 33 °C in ordinary godown and of maximum temperatures 20 °C and 40 °C. The relative humidity was between 60 and 90 per cent in the cool store and 15 to 65 per cent in the ordinary godown. Rotting percentage and driage in the store up to early June was generally lower in the former than in the latter. During storage, fewer eyes sprouted per tuber with Isopropyl-N-3-chlorophenyl carbamate (CIPC) treatment in the same period. Tetrachloro-nitrobenzene treatment had only a transitory effect on the number of eyes sprouted.

i) Evaporatively Cooled Potato Store (ECPS)

Attempts were made in India in the 1920s to store potatoes in an evaporatively cooled potato store. A 20 t capacity insulated potato store equipped with passive evaporative cooling, which does not require electricity or any other source of power, was evaluated at Central Potato Research Station, Jalandhar in 1981. It works on the principle of passive evaporation of water which results in the removal of heat from within the store lowering the inside temperature. It is a masonry structure measuring 9.1 x 4.6 x 3 m. The northern side has eight ventilators measuring 76 x 60 cm and the eastern wall has three ventilators, placed 15 cm above the ground level. The outer side of the ventilators is fitted with chicken wire mesh and 5 cm thick wood wool which is kept moist by water dripping from small holes made in a 1.25 cm water pipe placed above the ventilators and connected to a water tank. The ceiling is made of plywood mounted on wooden beams with 5 cm diameter holes at 10 x 10 cm spacing to facilitate the escape of hot air from within the store. The roof is made of galvanized iron sheet and is provided with four exhaust outlets (Figure 6). During the day the air inside the store becomes hot and rises through the ceiling and this is replaced by the fresh cool air entering through the moist wood wool which lowers the inside temperature. The temperature inside the ECPS ranges from 18 to 30 °C (6 to 12 °C lower than the ambient) and the RH is 85 to 95 per cent (15 to 42 per cent higher than the ambient). Due to low water vapor pressure deficit inside the ECPS, weight loss in potato tubers during storage is significantly reduced. Storage losses range from 10 to 15 per cent after 3 to 4 months of storage. Ware potatoes can be stored in good condition up to the beginning of June, when the prices in the market is much higher as compared to the prices at the time of harvest. Storage life of potatoes can be further extended up to the end of June by treating the tubers with a sprout suppressant like CIPC. Potatoes stored in ECPS have lower reducing sugar content of around 120 mg/100g tuber and produce highly acceptable light colour chips with colour scores of 2-3. Potatoes produced in north-western plains are otherwise unsuitable for processing due to high levels of reducing sugars at harvest. This store of 20 t capacity costs around Rs. one lakh and is suitable for regions where the temperatures are high and humidity is low during the storage period. The disadvantage with ECPS is that it works only during dry and hot summer and it cannot be used in rainy season. This structure can be used for only 3 to 4 months for the storage of potatoes and remains unutilized during the remaining 8 to 9 months. To make storage in ECPS economical, it needs to be used throughout the year.

5.1.2 Long-term Storage

Potatoes harvested in February-March have to be stored until October and cold storage is essential for such a long-term storage. The importance of potato storage in India was realized as early as 1925 and storing of seed potatoes from one crop to the next crop was considered the greatest constraint. One of the first cold stores for the storage of seed potatoes in the country was set up in U.P. in the year 1940. Growth of the cold storage industry was slow in 1940's and 1950's. Growth picked up in 1960's with the availability of technology and a large number of cold stores were built. Easy availability of compressors and insulating materials also contributed to that growth. With the increase in potato production in the country, cold storage business became profitable. Consequently competition started and large capacity cold stores (6-8 thousand metric tonnes) were constructed. This situation continued through 1970's. But in 1980's, there was little competition resulting in a monopoly like situation. Farmers found it difficult to get space in cold stores for the storage of potatoes. In the years of glut farmers faced very difficult situations and in some parts of the country, they had to reserve space in cold stores by paying the rent in advance. In the late 1990's, the cold storage industry went through hard times as electricity charges increased every year but rental rates did not increase proportionately. The situation changed in 2000 with the intervention of government through subsidy schemes for multi-product, multi-chamber cold stores. These schemes were also instrumental in introducing limited modernization with newly constructed cold stores switching over from diffuser system to fin coil. Many state governments removed rental controls. Beginning from the year 2000, the introduction of high speed compressors resulted in reduced power consumption and easy maintenance. New entrepreneurs are now more interested in building modern cold stores with improved insulation, and air circulation for better and uniform maintenance of temperature. With the construction of multi-chamber cold stores and storage of potatoes at 10-12 °C with CIPC treatment, there is increased awareness among the cold store owners about the importance of monitoring and maintenance of uniform temperature throughout the store, high level of relative humidity and low level of CO_2 concentration. The number of cold stores in the private cooperative and public sectors are 4885 (90.8 per cent), 356 (6.6 per cent) and 140 (2.6 per cent), respectively. Thus a look at the sector-wise distribution of cold stores reveals that cold stores in India are predominantly in the private sector (90.8 per cent in terms of number and 95.7 per cent in terms of capacity).Only a small percentage (2.6 per cent) are in the public sector. Data on the commodity-wise distribution of cold stores show that most of the available cold storage space is used for the storage of potatoes (53.2 per cent in terms of number and 75.4 per cent in terms of capacity). By the end of 2009, there were 2862 cold stores in the country with a capacity of 18.42 million tonnes, used exclusively for the storage of potatoes. This figure does not include multipurpose cold stores and those used for the storage of fruits and vegetables. About 77 per cent of the cold storage space available for potatoes is located in two states only *viz.* Uttar Pradesh (47.3 per cent) and West Bengal (29.6 per cent). Long-term storage of potatoes is expensive. The cost of storage is very high due to high cost of erection and maintenance of refrigerated storage structures, electricity, and loading and unloading. During storage, potatoes are affected by the storage temperature, humidity, composition of the storage atmosphere, their physiological condition at the time of storage and degree of microbial infection. Deterioration during storage is through weight loss (moisture loss), respiration, pest and disease attack and physiological disorders. All these factors have to be considered before deciding storage condition and duration. Storage of potatoes in cold store has become a common practice in the country.

Low temperature storage is the most common method of potato storage. Low temperature storage is effective because at storage temperature of 4 °C and below, sprout growth does not take place in most of the varieties. Further, the metabolic processes are slowed down therefore, losses caused by respiration are minimum. Pathogen metabolism is also slowed down at low temperatures hence rotting is checked. A normal cold store has a pre-cooling chamber where potatoes are kept at 15-16 °C for a few days and then shifted inside the cold store maintained at 2-4 °C and, 90-95 per cent RH. In India, cold stores were developed mainly for the storage of seed potatoes because non-refrigerated storage of seed potatoes in the first half of this century resulted in as high as 50 per cent storage losses. Storage requirements of seed potatoes is to minimize weight loss and suppress sprout growth without using a chemical sprout inhibitor so that the seed potatoes will be ready to sprout normally at the end of the storage period. But the present utilization pattern of cold storage space in different parts of the country shows that about 60 per cent of the cold storage space is being used for the storage of ware potatoes and 40 per cent for the storage of seed potatoes. Since the primary purpose of storage at 2-4 °C is to check sprout growth, storage of seed potatoes at 2-4 °C in cold store is ideal. But storage of ware potatoes at 2-4 °C is not desirable for the following reasons. i) Low temperatures induce accumulation of large amounts of sugars which make the tubers sweet and therefore less suitable for consumption. ii) Cold stored potatoes are unfit for processing as high level of reducing sugars causes browning in potato chips, iii) The keeping quality of cold stored potatoes deteriorates quickly once they are removed from the cold store as tubers begin to sprout, and iv) Marketing of cold stored potatoes is also a problem, as the trader cannot keep the potatoes at room temperature for long due to their poor keeping quality. The high sugar content of potatoes stored at 2-4 °C can be reduced to some extent by storing these potatoes at 20 °C for 6 weeks. This process is called reconditioning. However, reconditioning will not restore the sugar levels to those at the time of storage. Storage of processing potatoes requires a good technical knowledge of the crop.

Storage requirements of table or ware potatoes are to preserve their culinary properties, to prevent them from sprouting and to minimize storage losses. The problem of low temperature induced sweetness can be overcome by storing table and processing potatoes at 10-12 °C instead of 2-4 °C (Figures 7). However, the RH should be maintained at 90-95 per cent to minimize weight loss during storage. When stored at 10-12 °C, sugar accumulation is minimum therefore, potatoes do not taste sweet and hence are better than the cold stored potatoes for consumption. Since the sugar accumulation is minimum, they are quite suitable for processing, unlike potatoes stored at 2-4 °C. However, at10-12 °C, sprout growth takes place at the release of dormancy and therefore, CIPC a sprout suppressant is used to check sprout growth. Potatoes should be stored in leno bags to facilitate proper distribution of CIPC on the tubers during application.

As already mentioned, when potatoes are stored at 10-12 °C and 90 per cent RH sprout growth takes place. If sprout growth is not checked, it would result in excessive weight loss leading to shriveling of tubers. When tubers of variety Kufri Jyoti were stored at 10 °C, the weight loss was 14.2 per cent after 145 days of storage (DOS). But when sprout growth was checked with CIPC, the weight loss was 3.4 per cent after 145 DOS. The weight loss at 12 °C with CIPC treatment was 4.3 per cent after 145 DOS. When potatoes are stored at 10-12 °C with CIPC treatment, reducing sugar level is within the acceptable limit and chips produced from these potatoes are light in colour as compared to dark brown chips produced from cold stored potatoes. Storage of potatoes at 10-12 °C with CIPC treatment, on a commercial scale was started in India in 1998, when CIPC was registered for the first time for commercial use in India. Since then it has become popular in Indore and Ujjain in Madhya Pradesh. It is also being used in Meerut, Agra and Sambhal in Uttar Pradesh, Deesa in Gujarat and Sainthian in West Bengal.

Figure 3: Heap Storage of Potatoes.

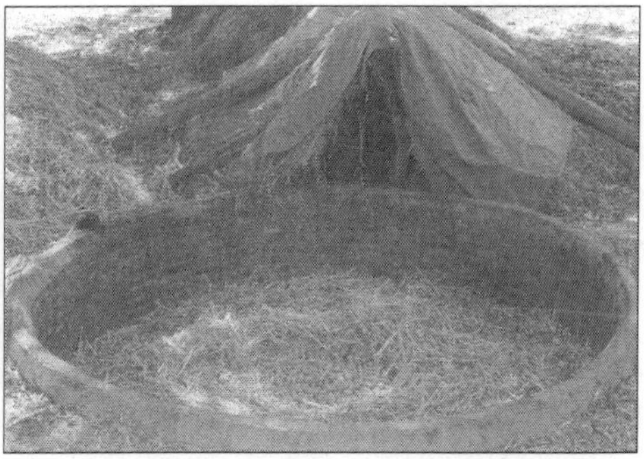

Figure 4: Pit Storage of Potatoes.

Inside view.

Figure 5: A Bamboo Structure for Potato Storage in Assam.

Figure 6: An Evaporatively Cooled Potato Store.

Figure 7: A Modern Cold Store with Chambers Used for Storage of Potatoes at 10-12 °C.

5.2 Quality of Stored Potatoes

As mentioned above, not all potato varieties can be stored at 10-12 °C with CIPC and used for processing. However, several of the varieties grown in the country can be used for table purpose after storage at 10-12 °C with CIPC. In a trial conducted at Central Potato Research Institute, Shimla, nine potato varieties grown at Modipuram *viz.* Kufri Jyoti, Kufri Chandramukhi, Kufri Sindhuri, Kufri Anand, Kufri Lalima, Kufri Bahar, Kufri Chipsona-1 and Kufri Chipsona-2, seven varieties grown at Jalandhar *viz.* Kufri Jyoti, Kufri Chandramukhi, Kufri Lauvkar, Kufri Badshah, Kufri Bahar, Kufri Sindhuri and Kufri Lalima, five varieties grown at Patna *viz.* Kufri Jyoti, Kufri Sindhuri, Kufri Ashoka, Kufri Chipsona-1 and Kufri Chipsona-2 and three varieties grown at Gwalior *viz.* Kufri Jyoti, Kufri Chandramukhi and Kufri Lauvkar were stored at 10-12 °C. CIPC treatment was given twice in the form of a fog. The formulation used was Oorja (Marketed by United Phosphorus Ltd., Mumbai) which contains 50 per cent a.i. It was applied @ 35 ml per t of potatoes. It was found that Kufri Jyoti, Kufri Lauvkar, Kufri Chandramukhi, Kufri Chipsona-1 and Kufri Chipsona-2 could be used for processing after storage at 10-12 °C with CIPC as the sugar level was within the acceptable limit. All the other varieties could be used for table purpose only. Though the sugar level in these varieties was higher than the acceptable level for processing, still they were much lower than that in potatoes stored at 2-4 °C.Thus, these potatoes were not sweet in taste and therefore, better than the cold stored potatoes for table purpose.

There is a wrong notion prevailing in some places that potatoes stored at 10-12 °C with CIPC treatment contain low level of reducing sugar therefore, they are better for diabetic patients. For this reason, some people even call these potatoes "diet potatoes." It is wrong to call these potatoes "diet potatoes" and to claim that these potatoes are better than other potatoes for diabetic patients. The reducing sugar content in potatoes stored at 10-12 °C can vary from less than 100 to 300 mg/100g f. wt, whereas the reducing sugar content in cold stored potatoes can vary from less than 1000 to 3000 mg/100g f. wt. Because of the difference in storage temperatures, the sugar content is lower and starch content is higher in potatoes stored at 10-12 °C as compared to potatoes stored at 2-4 °C, which contain higher reducing sugar and lower level of starch. When the diet of diabetic patients is discussed, not only the free sugar level should be considered but also the total carbohydrate content of the food is to be considered. Therefore, if we consider the total carbohydrate content, there is little difference between cold stored potatoes stored at 2-4 °C and those stored at 10-12 °C with CIPC treatment. Further, the reducing sugar content in potatoes stored at 10-12 °C is comparable to freshly harvested potatoes. Therefore, they are as good as freshly harvested potatoes in taste. Nutrition wise barring the small loss in free sugars which are used up in respiration during storage there is not much difference in the carbohydrate level among fresh potatoes, potatoes stored at 10-12 °C with CIPC treatment and cold stored (2-4 °C) potatoes. All these potatoes can be consumed by diabetic patients provided the total carbohydrate intake does not exceed the level recommended by the doctor. Several studies carried out in Europe have clearly shown that potatoes do not cause or worsen diabetes.

5.3 Consumer Preference for Stored Potatoes

Surveys were carried out in five cities *viz.* Meerut (UP), Deesa (Gujarat), Indore (MP), Ujjain (MP) and Sainthian (WB), to see the preference of consumers to CIPC treated potatoes stored at 10-12 °C for table purpose in relation to cold stored (2-4 °C) potatoes which are normally available in the plains from May onwards. When asked which potatoes had better appearance, 61 per cent of the consumers felt cold stored potatoes had better appearance. When asked which potatoes they found sweet in taste, 90 per cent of the consumers said they found cold stored potatoes sweet. In Deesa and Sainthian, 100 per cent of the consumers felt cold stored potatoes were sweet. Majority of the consumers (58 per cent) preferred CIPC treated potatoes because they were not sweet in taste. Although CIPC treated potatoes were preferred for their better taste, not everyone who preferred them were willing to pay more for them. Only 42 per cent were willing to pay more for CIPC treated potatoes. Two cases *viz.* Meerut and Deesa make an interesting study. While 91 per cent of consumers in Meerut preferred potatoes stored at 10-12 °C with CIPC, 96 per cent of consumers in Deesa preferred potatoes stored in cold store at 2-4 °C (Table 4). This is because, consumers in Deesa like sweet taste in their vegetable preparations.

Table 4: Consumer Preference for CIPC Treated Potatoes.

Particulars	*Meerut (UP)*	*Deesa (Gujarat)*	*Indore (MP)*	*Ujjain (MP)*	*Sainthian (WB)*	*Mean*
Sample Size	*57*	*69*	*54*	*73*	*29*	
Preference for appearance						
i) CIPC treated (10-12C)	59.6	11.6	35.2	43.8	44.5	38.9
ii) Cold stored (2-4C)	40.4	88.4	64.8	56.2	55.5	61.1
Which potatoes were sweet						
i) CIPC treated (10-12C)	1.8	0	25.9	21.9	0	9.9
ii) Cold stored (2-4C)	98.2	100	74.1	78.1	100	90.1
Preference for taste						
i) CIPC treated (10-12C)	91.2	4.4	51.8	75.3	65.5	57.6
ii) Cold stored (2-4C)	8.8	95.6	48.2	24.7	34.5	42.4
Willingness to pay more for CIPC treated potatoes						
i) Yes	54.4	4.4	40.7	46.5	62.1	41.6
ii) No	45.6	95.6	59.3	53.5	37.9	58.4

Values given are in percentage.

5.4 Biochemical Changes during Storage

The metabolic processes remain active in tubers even after their detachment from the plant. Storage conditions can alter the normal metabolic processes in a potato tuber. Major biochemical changes that occur in potatoes during storage are briefly discussed here.

5.4.1 Carbohydrates

Carbohydrates changes in potatoes during storage have been investigated in great detail. Only those changes that have a direct bearing on processing quality of potatoes are discussed here. Carbohydrates constitute 75 to 85 per cent of potato tuber dry matter, most of which is starch. The carbohydrate content and composition are important with respect to potato processing and its nutritive value. During storage, there is a gradual loss of carbohydrates from potatoes due to respiration. The rate of respiration is higher at temperatures below 4 °C, increases gradually up to 15 °C and then steeply with further increase in temperature. It has been observed that the respiratory peak corresponds with the peak of sucrose concentration. At low temperatures, therefore, besides increased respiration, free sugars also accumulate to cause low temperature-sweetening. Potatoes stored at low temperatures thus become unsuitable for processing into chips or French fries. The correlation between rate of respiration and sucrose content can be explained as both result from alternative metabolic pathways of same substrate glucose-1 phosphate, produced by the phosphorylysis of starch, the ultimate respiratory substrate in potato tuber. The low temperature sweetening is negligible at around 10 °C. However, sugar content can increase at 10 °C also if sprouting takes place and the mobilized sugar is transported into the young sprouts. To prevent sugar accumulation at 10 °C, sprouting should be checked with a sprout suppressant like CIPC. Gamma irradiation has been reported to increase the sucrose content of potato tubers. Potato storage at higher temperatures ranging from 24.7 to 36.2 °C results in increase in total sugars but the increase is much smaller than that in cold storage. At higher storage temperatures, loss of dry matter and rotting increase sharply.

5.4.2 Nitrogen Fractions

The total nitrogen of potato tuber comprises soluble protein, insoluble protein and soluble non-protein nitrogenous compounds (NPN). The NPN compounds are free amino acids, the amides asparagines and glutamine and small amounts of nitrate nitrogen and basic nitrogen compounds including nucleic acids and alkaloids. Most of the storage studies have been carried out to investigate changes in free amino acids, soluble proteins and glycoalkaloids. The breakdown of proteins to free amino acids during prolonged storage of potatoes has been reported. Though reducing sugars are the major determinants of processing quality, anomalies in the relationship of reducing sugars to fry colour do occur and free amino acids may also act as determinants of processing quality, as illustrated by the greater browning per unit reducing sugar in tubers grown under high nitrogen conditions. Protein degradation has been associated with the end of tuber dormancy and the mobilization of nitrogen reserves for sprout formation and the breakdown of proteins is consistent with an increase of proteolytic enzyme activity. Protein synthesis may also occur during potato storage, although low temperature conditions reduce the rate of this process while enhancing protease activity. Storage studies at 5 and 10 °C up to 40 weeks have shown that free amino acids content of stored potato tubers is controlled by a combination of factors including proteinase activity and protein synthesis, the overall balance of nitrogen flow depending on the dormancy status of the tubers. Consequently free amino acid content changes with storage duration. Glycoalkaloids are another important nitrogen fraction and about 95 per cent of the total glycoalkaloids present in potato tubers are accounted for by two compounds, *viz.* alpha solanine and alpha chaconine. It has been suggested that at low concentrations the presence of glycoalkaloids may enhance the flavour of potatoes, but at levels exceeding 15 mg/ 100g FW, a bitter taste may be perceived. Potatoes containing more than 20 mg of glycoalkaloids per 100g FW of tuber are not safe for human consumption. When potatoes were stored at 4, 7 and 10 °C for 9 weeks it was observed that at all the three storage temperatures, the rate of glycoalkaloid accumulation was independent of the levels found at harvest and significant interactions were found between the cultivars and temperature.

5.4.3 Enzyme Systems

The potato tuber contains numerous enzyme systems that constitute a considerable proportion of the total protein, and some of which are important in relation to its processing and nutritional quality. Most of the studies carried out on this aspect deal with the enzymes associated with low temperature sweetening. Starch degrading enzymes decrease the degree of polymerization of starch, which contains predominantly alpha 1, 4-linked glucose units. The process of starch degradation during cold sweetening is believed to be principally phosphorylitic rather than hydrolytic in nature. In potato tubers two starch phosphorylase enzymes, types L and H, have been described and are believed to be responsible for the complete breakdown of starch. Some studies have suggested that an increase in starch phosphorylase activity acts as a triggering event in the sweetening of potato tubers during cold storage. Studies involving alpha amylase, beta amylase and debranching enzymes in potatoes stored at 4 and 10 °C showed that activities of these three enzymes and reducing sugar content increased at 4 °C while only a slight increase was observed at 10 °C. A cold inducible beta amylase has been purified and characterized from the variety Desiree. Its activity was low in tubers stored at 20 °C and was progressively increased when temperatures were decreased to 5 and 3 °C. The studies on cold resistant potato clone ND 860-2 suggest the ordered crystalline structure of starch granules as a key factor for its reduced sweetening potential in cold storage. Sucrose phosphate synthase, UDP glucose pyrophosphorylase (UGPase) and invertase are the three major enzymes directly involved in sucrose and reducing sugar formation. These enzymes have also been investigated in detail with respect to low temperature sweetening. In cold stored tubers, 70-80 per cent decrease in sucrose phosphate synthase expression resulted in a reproducible but non-proportional (10-40 per cent) decrease of soluble sugars. A relationship has been shown to exist between the allelic polymorphism of UGP ase exhibited among potato clones and their ability to accumulate sugars in cold storage. Genetic modification of acid invertase activity in potatoes has resulted in decreased concentrations of reducing sugars. Eight Indian potato varieties were stored at 3-5 °C and reducing sugars and invertase activity was monitored at periodic intervals. Reducing sugars and both basal and total invertase activity increased during storage in all the varieties, and there was a significant correlation between reducing sugar content and basal and total invertase activity.It was initially theorized that the cold liability of glycolytic enzymes led to an increase in hexose phosphate concentrations resulting in a stimulation of sucrose synthesis. But results of plant transformation studies have shown that there is no evidence to support the theory that a cold labile step in glycolysis is involved in triggering or regulating cold induced sugar accumulation in potatoes.

5.4.4 Other Biochemical Changes

Potatoes are a major and cheap source of ascorbic acid (vitamin C) in the human diet, but its levels decrease during storage. The most rapid loss occurs during the first months of refrigerated storage ranging from 40 to 60 per cent. Storage temperatures as well as time affect vitamin C losses and the losses are higher below 10 °C storage than between 10 and 20 °C. Polyphenolic compounds are present in potatoes and enzyme catalysed browning reactions involve the oxidation of phenolic compounds to quinnones, followed by transformation of the quinines to dark pigmens by the enzyme tyrosinase and polyphenol oxidase (PPO). These changes result in deterioration in flavour, colour and nutritional quality. The storage of potatoes at high temperatures did not affect the total phenolic and chlorogenic acid content, but both levels increased on storage at 10 °C for five weeks. PPO activity is higher at higher storage temperatures leading to greater discolouration. The principle organic acids in potatoes are citric acid and malic acid. Low temperature storage resulted in decrease in pH and an increase in malic and citric acid contents of tubers.

5.5 Economics of Storage

During the period of a study on economics of potato storage, the normal cold storage rent varied from Rs. 0.60-0.65 per kg whereas, for CIPC treated potatoes, the rent charged varied from Rs. 1.0 –1.16 per kg. The most common modifications required for using an existing cold store for storing potatoes at 10-12 °C with CIPC treatment are as follows. (i) Partitioning is normally required as the existing chamber dimensions use to be too big for CIPC treatment (ii) Ducts may have to be installed to facilitate flushing out of accumulated carbon dioxide (iii) Humidifiers will be needed to maintain the R.H. The additional expenditure which cold store owners incur in making these modifications ranged from Rs. 0.01 to 0.11 per kg of potatoes. The cost of CIPC treatment varied from Rs. 0.11 to 0.36 per kg of potatoes. Such large variation in the cost of treatment was observed because some cold store owners used higher dosage of CIPC to account for leakage during treatment. While several cold store owners were able to get additional profit ranging from Rs. 0.04 to 0.37 per kg of potatoes, some cold store owners incurred losses due to various reasons like increased rottage and fall in potato prices at the time of selling. Farmers and traders who stored their potatoes at 10-12 °C with CIPC treatment were able to increase their profit by 37 to 58 per cent in some places but in other places farmers and traders incurred losses because there was a fall in potato prices at the time of selling and as a result, the cost of storage was higher than the rise in sale price of potatoes after storage. Only those, whose potatoes were acceptable to the processors, were able to make profit. But in many cases, the potatoes were not acceptable to processors, therefore, had to be sold as table potatoes resulting in either marginal profit or loss depending upon the market situation.

5.6 Future Storage Needs

The distribution of cold stores in the country is uneven. Most of the cold stores are concentrated in a few regions only. There is a need to identify areas that lack adequate cold storage facilities and promote the setting up of new cold stores in such areas in the future. Construction of new cold stores should be permitted based on potato production and seed requirement of a given area to increase accessibility to farmers. Cold stores were developed primarily to store seed potatoes at 2-4 °C. But at present cold stores maintained at 2-4 °C are used for the storage of seed as well as table potatoes. Storage at 2-4 °C requires considerable energy. Cold stores maintained at 2-4 °C should be used only for the storage of seed potatoes. If 60 per cent of cold storage (2-4 °C) space is converted into storage space for processing and ware potatoes to be stored at 10-12 °C, it will result in considerable saving of electricity. New cold stores should be encouraged to maintain two temperatures *i.e.* 40 per cent storage space at 2-4 °C to meet the storage requirements of seed potatoes and 60 per cent storage space at 10-12 °C to meet the storage requirements of ware and processing potatoes.

Majority of the existing cold stores have difficulty in maintaining uniform temperature, relative humidity and appropriate air circulation. The fresh air requirements of potatoes are not being met. However, awareness is increasing and many cold store owners are introducing new technologies to improve uniform maintenance of temperature through better air circulation, and to increase RH levels in the cold stores by using humidifiers. Cold store owners storing potatoes at 10-12 °C with CIPC treatment have also started monitoring CO_2 concentration in their cold stores and flush out accumulated CO_2 frequently. There is plenty of scope for improvement and modernization. Monitoring of temperature, relative humidity and CO_2 concentration on a continuous basis using computer controlled monitoring systems should become an integral part of every cold store. Arrangements should be made for improved availability of fresh air to the potatoes to meet their respiratory demand and for uniform air circulation to avoid development of hot pockets within the cold store. Humidifiers need to be used for maintaining high level of RH in the storage environment. Due attention should be paid to aspects like proper curing of potatoes before storage, and loading rate and pull down time at the time of storing the potatoes. Details on storage conditions required for optimum storage of Indian Potato varieties grown under different agro climatic conditions need to be developed. The projected target for potato production in 2012-13 is about 40 million t. If storage of about 66 per cent of the total production is targeted, cold storage space will be required for about 26 million t of potatoes. As per a projection of National Horticulture Board, the additional cold storage capacity required is 8.280 million t and that will require an estimated investment of Rs. 66,240 million. It is hoped that this additional cold storage space to be created in the near future will make use of the latest technology available for the most efficient way of storing potatoes for up to 10 months with minimum storage losses, and with no deterioration in the quality of stored potatoes.

5.7 Use of Chemicals in Potato Storage

5.7.1 Sprout Inhibitors

Once the dormancy is released, sprouts begin to grow. Potatoes meant for ware and processing use are generally stored at 10-12 °C and at these storage temperatures sprout growth takes place which need to be checked with a sprout inhibitor. Sprout growth takes place in potatoes stored under non-refrigerated conditions also and use of a sprout inhibitor is needed to check sprout growth. Several sprout inhibitors have been used on potatoes. Some of the compounds used for checking sprout growth included several alcohols, acetaldehyde, and ethylene. Naphthalene acetic acid (NAA) and its derivative methyl ester of alpha naphthalene acetic acid (MENA), 2-4 dichlorophenoxy acetic acid (2, 4-D) and 2, 4, 5-trichlorophenoxy acetic acid (2, 4, 5-T) have also been reported to show sprout suppression effect. Although a number of chemical sprout inhibitors have been tried, only three have gained commercial acceptançe. They are tetrachloro nitrobenzene, maleic hydrazide and isopropyl-N-chlorophenyl carbamate (CIPC). Of these, only CIPC is used commercially in India.

a) Maleic Hydrazide (MH)

Maleic Hydrazide is a sprout inhibitor that is applied as a pre-harvest spray to the foliage of potato crop. Work done at CPRI has shown that glyphosate can also be used as a sprout inhibitor that is applied as a pre-harvest spray to the foliage. Maleic Hydrazide is a sprout inhibitor that is applied as a pre-harvest spray to the foliage. MH applied to the foliage is translocated to the tubers and 10-15 days are needed for the accumulation of MH in tubers in sufficient concentration to inhibit sprout growth. MH was the only sprout suppressant registered for commercial use in India until 1998. 0.30 per cent MH applied 4 weeks before haulm cutting is effective in suppressing sprout growth without any adverse effect on tuber yield and quality. However, the time of MH application is crucial. If it is applied early yield is reduced, and if it is applied late, it is less effective. Because of the difficulty in applying at the right time, it is not being used on potatoes in India. Furthermore, often sprout suppression is not complete in MH treated tubers, and growth of cauliflower like sprouts has been observed. The tolerance limit to MH residues is 50 ppm. MH is not in use in European countries also. However, as per latest notification of the Central Insecticide Board, Govt. of India dated 1st January, 2014 manufacture, import and use of Maleic Hydrazide (MH) is banned in India.

b) Glyphosate

Glyphosate [*N*-(phosphonomethyl) glycine] is another sprout inhibitor that can be applied as a pre-harvest spray to foliage. It is considered a non-residual and environmentally benign herbicide. Around 20 years back it was observed by researchers that when glyphosate is applied as a preharvest foliar spray, it affects the yield of potato. However, further work indicated that at lower doses it does not affect the yield but, has a strong effect on the sprout growth of the potatoes produced by the treated plants. Glyphosate applied to the foliage is translocated to the tubers where it affects sprout growth by inhibiting meristematic development of sprouts. EPSPS (5-enolpyruvylshikimate 3-phosphate synthase) is a key enzyme of the shikimate pathway which is required in plants for the synthesis of aromatic amino acids and other compounds including vitamins, plant growth substance and lignin and this enzyme is the target of glyphosate. In view of the versatile role of EPSPS enzyme in the plant metabolism, its inhibition by glyphosate affects many physicochemical and physiological processes in plant. Work done at CPRI, Shimla has indicated that single preharvest foliar application of glyphosate at sub-lethal dose of 0.694 kg (a.i.) in 600 litres of water ha^{-1} at 60 days after planting is effective in suppressing the sprout growth of the harvested potatoes during storage at 18 °C However, more research work is needed on glyphosate for establishing it as an effective sprout suppressant on potato.

5.7.2 Postharvest Application

CIPC and TCNB are the two sprout inhibitors applied postharvest to potatoes in storage

a) Isopropyl-N-3-Chlorophenyl Carbamate (CIPC)

CIPC is the most commonly used sprout suppressant on potatoes in many countries. It is very effective in checking sprout growth, when applied as a postharvest treatment on potatoes during storage at 10-12 °C. Details on this chemical, its application and effectiveness have already been mentioned above.

b) CIPC

CIPC (isopropyl N (3-chlorophenyl) carbamate) is the most common sprout suppressant used on potatoes. It is commonly called chlorpropham. CIPC is a mitotic inhibitor and checks sprout growth by inhibiting cell division. It inhibits spindle formation during cell division. CIPC was registered for use in India in 1998, and since then several cold stores in the country are using this chemical on potatoes stored at 10-12 °C. CIPC is available in three formulations, *viz.* solution form to be applied as an aerosol, emulsifiable concentrate for use as a direct spray and powder form for direct application. The liquid formulation is available at concentrations ranging from 25 to 50 per cent a.i. (active ingredient) and a fogger is used for its application. The rate and dose of application depends on the concentration of active ingredient, variety, duration of storage and temperature of storage. The emulsifiable concentrate is generally used on export consignment. The CIPC powder or dust is normally available as 1 per cent or 2 per cent active ingredient and is applied through applicators on potatoes going into store through conveyor belts. Powder applications could affect the natural skin colour of potatoes. A good storage can only maintain the quality of potatoes stored but it cannot improve it. Therefore, it is essential that potatoes for treatment with CIPC should be physically and chemically mature. The physical maturity refers to proper skin curing and wound healing. Potatoes treated with CIPC cannot form periderm layer as cell division is inhibited. Chemical maturity refers to low reducing sugars and sucrose levels. High sucrose content at the time of storage can lead to accumulation of reducing sugars during long term storage.

CIPC Treatment

Table potatoes can be treated with 25 mg CIPC dust a.i./t of tubers (commercial preparation of CIPC dust contains 1 per cent active ingredient) at the time of storage. For treating small quantity of potatoes, powder form of CIPC is convenient but for treating large quantity of potatoes, powder form is not convenient due to the practical problems in handling, since powder treatment is done manually in India. For large scale treatment of potatoes with CIPC, liquid formulation is preferred as it can be applied in the form of a fog using a fogging machine, without disturbing the stored potatoes. Chlorpropham (CIPC) is being marketed in India by United Phosphorus Ltd., Mumbai in the name of Oorja. This commercial preparation contains 50 per cent a.i. and 30-35 ml of this preparation is required for fogging one t of potatoes. However, this dose can vary depending upon the loss of fog during treatment due to leakage, the pore size of the bags used, the quantity of potatoes stored etc. After the CIPC treatment, the store should be kept air tight for 24-48 hours. The treated potatoes are safe for consumption after 3 to 4 weeks of treatment. First fogging is done at the first sign of sprout growth. Second fogging is done at about 45 days after the first fogging. The timing of first fogging is important. If it is done at the right time, CIPC is most effective in suppressing sprout growth (Figure 8). Dealyed treatment of CIPC at inadequate concentration and after sprouting make it less effective and there is the risk of internal sprout growth. Figure 9 shows a tuber showing internal sprout growth after inadequate CIPC application. When internal sprout growth takes place, tubers become useless and have to be discarded. After CIPC application, ventilation should not be done for 24 hours. After this, regular flushing out of excess carbon dioxide is essential to maintain the quality of potatoes. Prolonged storage at 10-12 °C could result in sweetening, which is variety dependent. Senescent sweetening is irreversible and sugar level cannot be decreased through reconditioning. Contamination of seed potatoes with CIPC should be prevented as this can impair the emergence of the crop.

(a) Untreated Control **(b) CIPC treated**

Figure 8: Effect of CIPC on Sprout Growth of Potatoes Stored at 10-12 °C.

CIPC Residue Level

Residue level of chemicals in treated foods has been a health concern. Therefore, it is important that we know what the residue level of CIPC in treated potatoes is. Tuber samples were collected from two cold stores each at Indore, Ujjain and Meerut at regular intervals beginning from two days after treatment and CIPC residue levels in peels of treated potatoes and in unpeeled potatoes were determined. The maximum residue level observed in peels was 14.7 ppm and the maximum residue level observed in unpeeled tuber was 2.5 ppm. A tolerance limit of 30 mg/kg tubers (30 ppm) has been given by EPA. Since we did not observe more than 2.5 ppm in unpeeled tubers, it can be said that CIPC treated tubers are safe for human consumption. To be on the safer side, it is recommended that CIPC treated tubers should be consumed only after 3- 4 weeks of treatment.

c) Tetrachloro-Nitrobenzene (TCNB)

TCNB also known as Tecnazene (Commercial name 'Fusarex') is in commercial use in some countries but its use is limited. It is known to suppress sprout growth, when given as postharvest application on potatoes stored at 10 °C and does not inhibit wound healing. Since it is highly volatile, it can be used on seed potatoes also. It is applied at the rate of 135 mg/kg of potatoes.

d) Natural Substances as Sprout Inhibitors

The growing tendency to avoid consuming foods treated with chemicals has led to increasing research interest in natural substances with sprout inhibitory effect. The ultimate aim is to replace chemical sprout inhibitors with naturally occurring compounds. Several naturally occurring aromatic aldehydes (salicylaldehyde, benzaldehyde, cinnamaldehyde and cumminaldehyde) and aromatic alcohols (thymol) are reported to inhibit sprout growth for a short period of up to 14 days. Aromatic compounds like mono, di and trimethyl naphthalenes, benzothiophene and diphenylamine have also been found to have sprout inhibitory effect. Some volatile monoterpenes are also reported to inhibit sprout growth. Muna plants (family *Lamiacea*) contain essential oils that are comprised of over 98 per cent monoterpenes. Small scale experiments have shown that pulegium containing essential oils of *Minthostachys glabrescens* and *Mentha pulegium* and the carvone containing essential oils of *Anathem graveolens* and *Carum carvi* are efficient sprout inhibitors. 1, 8-cineole, available in large quantities (70-85 per cent) in oil of *Eucalyptus globules*, is a promising monoterpene with good sprout suppression effect. Vapour treatment (250 µl/l for ten days) with pure *E. globules* oil showed strong sprout suppression effect up to 80 days of storage at 24+ 2 °C. Carvone is the only monoterpene produced by *Carum carvi* (caraway), which has shown consistently good results in storage trials at 8-10 °C. It is found naturally in essential oils of *Cuminum cyminum, Mentha spicuta, Mentha citrata, Mentha cardiac, Anathem graveolens* and *Anathem soya*. The S-carvone form is being marketed in the Netherlands under the commercial name "Talent". Since carvone is highly volatile, repeated applications are necessary to check sprout growth. This makes it very expensive and carvone treatment may be 10 times more costly than CIPC treatment.

5.7.3 Irradiation

Irradiation has also been used to check sprout growth in potatoes. Irradiation suppresses sprout growth by preventing cell division. In the early 1960's, gamma irradiation was tried to inhibit sprout growth in potatoes of old varieties like Phulwa and up-to-date. Periderm formation in stored potatoes is completely inhibited even when a low dose of 20-30 Gy gamma irradiation is used, as it affects the nuclei of potential periderm cells. Considerable amount of work on the use of gamma irradiation for sprout suppression of potatoes stored under various storage conditions has been done by BARC, Mumbai and CPRI, Shimla. These trials led to the conclusion that gamma irradiation cannot be used for suppressing sprout growth in potatoes stored under non-refrigerated conditions as higher temperatures during storage resulted in unacceptably high storage losses in irradiated potatoes due to increased rotting. Rotting generally increases in irradiated potatoes as irradiation interferes with wound healing by inhibiting cell division. Trials conducted at BARC have shown that under refrigerated conditions, irradiated potatoes could be stored successfully up to nine months at 10 °C. In a trial conducted at CPRI, Shimla during 2003-04, potatoes of three varieties *viz.* Kufri Chandramukhi, Kufri Jyoti and Kufri Chipsona-2 were exposed to 0.1 kGy and 0.5 kGy gamma irradiation and stored at 8, 12 and 16 °C. Potatoes could be stored for six months at these three temperatures but rotting was on the higher side in all the three varieties. While 0.1 kGy gamma irradiation was as good as the chemical sprout suppressant CIPC in suppressing sprout growth of potatoes (Figures 15a&b), irradiation induced accumulation of sugars made the stored potatoes less suitable for processing. Irradiation caused 18-35 per cent increase in reducing sugars content and 20-29 per cent increase in sucrose content. Higher concentration of reducing sugars resulted in browning of fried potato chips (Figures 15a&b). However, chemical sprout suppressants are preferred to irradiation, which has not found wide acceptance globally.

6.0 Conclusion

Potato production in the country has shown a phenomenal growth. The steady increase in the total production at the national level has not been matched by the developments in postharvest technology. This mismatch between potato production and postharvest technology and utilization has resulted in periodical gluts. Adoption of alternate storage methods for short-term storage of potatoes will reduce the demand for cold storage space in the country. There are a number of traditional, low-cost and non-refrigerated storage structures (essentially based on evaporative or passive evaporative cooling) in use in India with varying degrees of success. The traditional methods have been studied, validated and popularized for particular regions. In non-refrigerated storages, use of sprout suppressants has also been popularized to prevent excessive weight loss and shrinkage due to sprouting. Storage losses in traditional methods like heaps and pits could be quite high if proper care is not taken. For these methods of storage to become more widely acceptable throughout the country, it is necessary to improve their efficiency and to reduce the storage losses. Potatoes stored in heaps and pits are suitable for processing due to their higher dry matter and lower reducing sugars content. Storage of potatoes in cold store has become a common practice in the country and potatoes are generally stored in cold store at 2-4 °C and 90-95 per cent RH. When stored at this temperature potatoes do not sprout and weight loss is minimum. The conditions maintained in cold stores are ideal for storing seed potatoes but not for the storage of table and processing potatoes. Cold stored potatoes are especially unsuitable for processing since higher sugar levels result in unacceptable, dark brown chips. The problem of low temperature induced sweetness can be overcome by storing table and processing potatoes at 10-12 °C instead of 2-4 °C.Storage of potatoes at 10-12 °C with CIPC treatment is becoming popular in India. However, the storage technique being practised at present is not satisfactory. Greater awareness about suitability of a potato variety for storage at 10-12 °C, maximum duration of storage for a given variety at this temperature without affecting the processing quality and right time of application of CIPC, will help in getting the maximum benefit. Farmers, traders and cold store owners can gain from this storage technology if the processing quality of potatoes stored is acceptable to potato processors in the country. There is plenty of scope for improvement and modernization. Monitoring of temperature, relative humidity and CO_2 concentration on a continuous basis using computer controlled monitoring systems should become an integral part of every cold store. Potato cold storage industry has played a very significant role in sustaining the increased potato production in the country during the last sixty years. However, there are problems that need to be addressed. Unless immediate attention is paid to the problems associated with potato storage and utilization, it will not be possible to sustain the current trend of increase in potato production.

7.0 Important References

Burton, W.G. 1989. The potato, 2nd Edn., Longman, Essex, 742 pp.

Burton, W.G., A Van Es and K.J. Hartmans, 1992. The physics and physiology of storage. In: The potato crop, 2nd Edn. (P.M. Harris, Ed.), pp. 6-8-727. Chapman and Hall, London.

Gottschalk, K.and Ezekiel, R. 2006. Storage. In: Handbook of potato production, improvement and postharvest management. (J.Gopal and S.M.P.Khurana, Eds), pp 489-521, The Haworth Press, New York.

Sukumaran N.P. and S.C.Verma 1993. Storage and processing. In: Advance in horticulture. Vol. 7-Potato (K.L. Chadha and J.S. Grewal, Eds), pp. 701-732, Malhotra Publishing House, New Delhi.

Talburt, W.F. and O. Smith. 1987. Potato processing. 4th Edn. Van Nostrand Reinhold, New York. 796 pp.

http://www.potaoworld.in

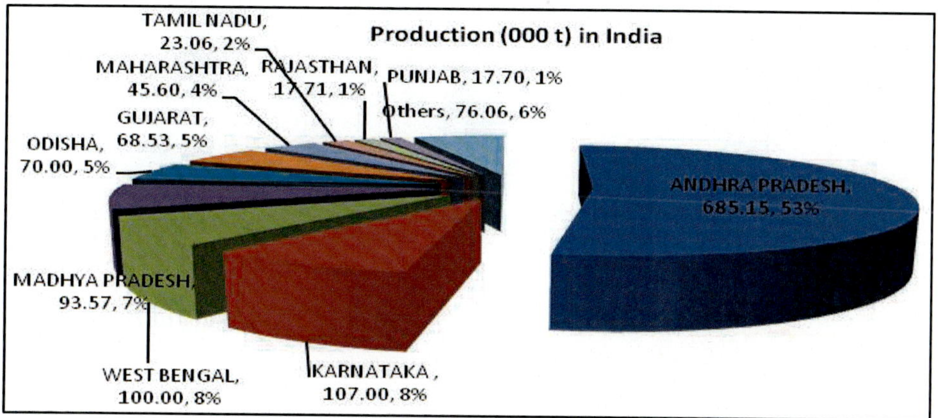

Dry Chilliy Production and Production of Share of different States in India, 2012 (p. 599)

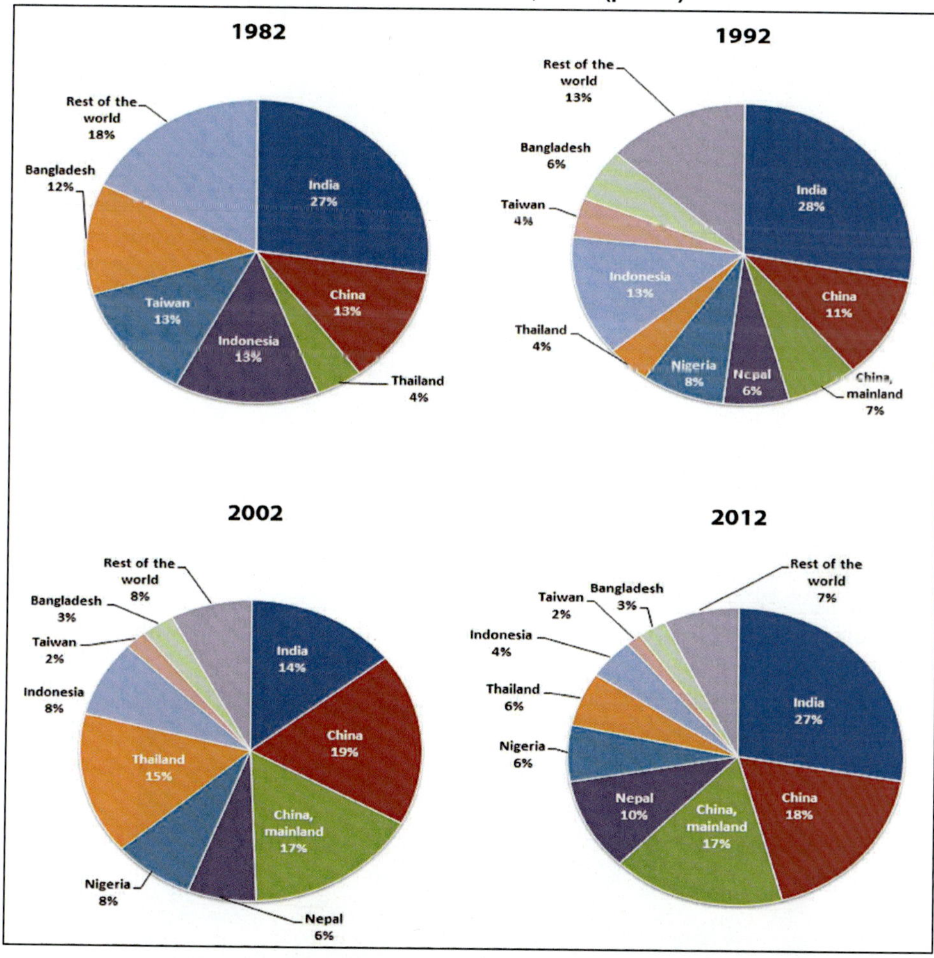

World Productivity of Ginger between 1982 to 2012 (FAO) (p. 657)

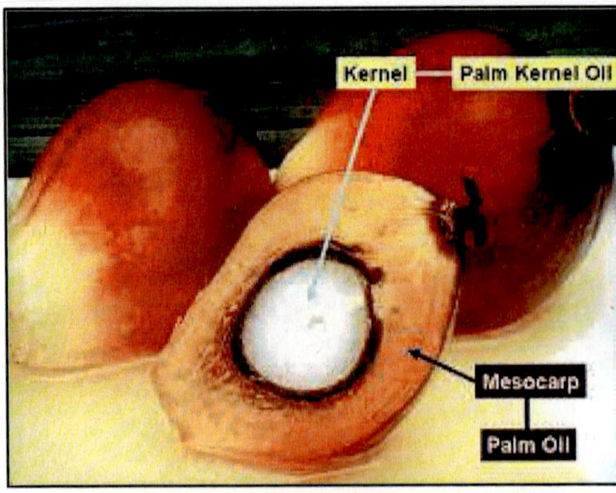

Figure 1: Kernel of Oil Palm (p. 686)

Onion Grader (p.700)

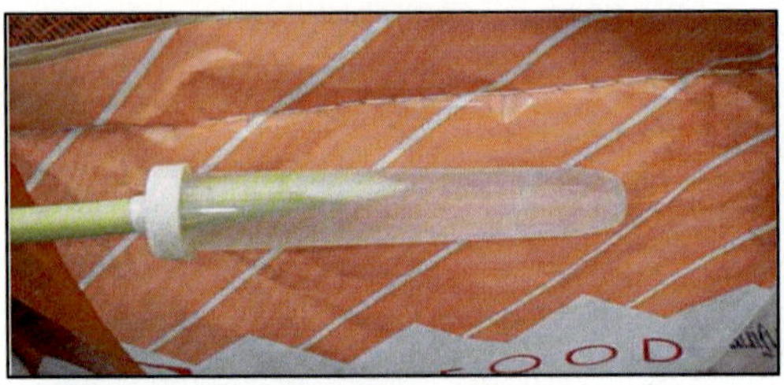

(a) Cut Flowers are Inserted in Tube with

(b) Individual Spikes Inside the Box in Alternate Fashion for Domestic

(c) Polyethylene Wrapping

(d) Packing of Cut Cymbidiums with Local Packaging

(e) Packing of Cut Cymbidiums for Export Markets

Packaging of Cut Flowers (p.721)

***Cymbidium*, 'Fire Storm Ruby'**

***Dendrobium*, 'Dang Saard'**

***Oncidium*, 'Sharry Baby Sweet Fragrance'**

***Vanda*, 'Roberts Delight Blue'**

***Phalaenopsis*, 'Strawberry'**

Commercial Orchids Varieties (p.714)

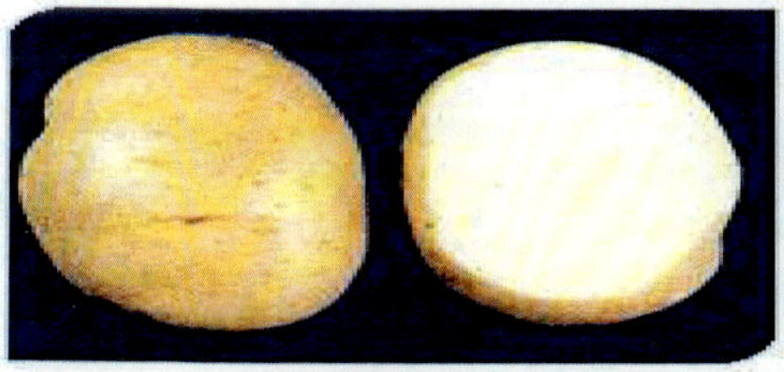

(a) Kufri Chandramukhi

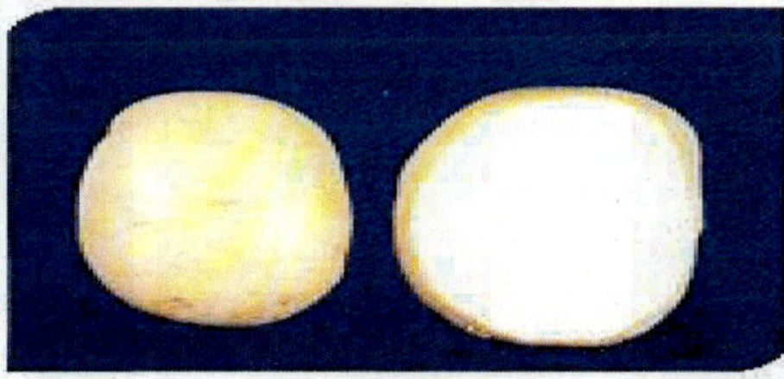

(b) Kufri Chipsona-I

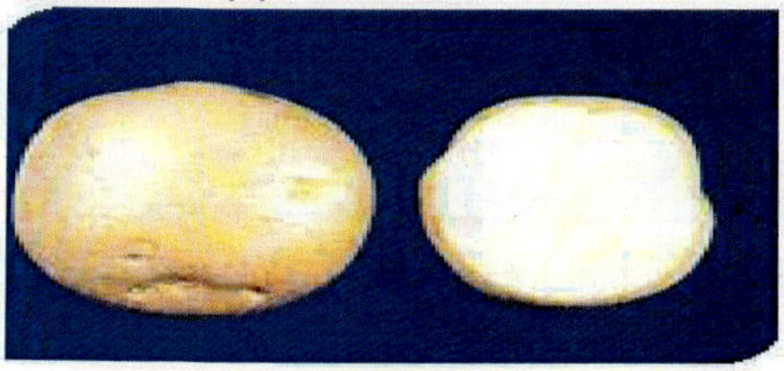

(c) Kufri Jyoti

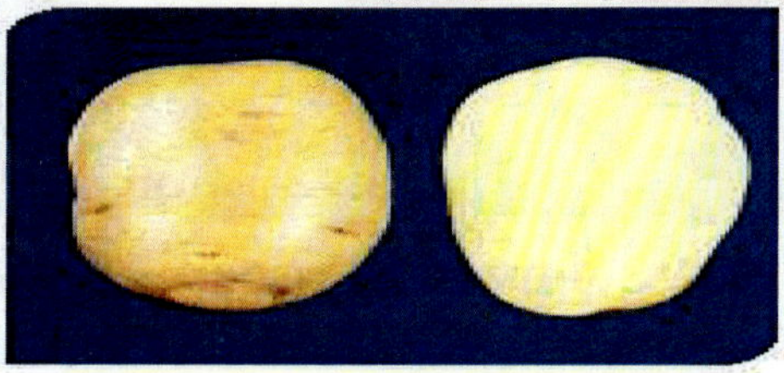

(d) Kufri Pukhraj

Commercial Cultivars of Potato (p.730)

(a) Kufri Chandramukhi

(b) Kufri Chipsona-I

(c) Kufri Jyoti

(d) Kufri Pukhraj

Commercial Cultivars of Potato (p.730)

(a) Immature Tubers and (b) Well Cured Mature Tubers (p. 731)